Community Seed Production Sustainability in Rice–Wheat Farming

Narayan Prasad Khanal • Keshav Lall Maharjan

Community Seed Production Sustainability in Rice–Wheat Farming

Springer

Narayan Prasad Khanal
Hiroshima University
Higashi-Hiroshima, Hiroshima
Japan

Keshav Lall Maharjan
Hiroshima University
Higashi-Hiroshima, Hiroshima
Japan

ISBN 978-4-431-55473-8 ISBN 978-4-431-55474-5 (eBook)
DOI 10.1007/978-4-431-55474-5

Library of Congress Control Number: 2015938915

Springer Tokyo Heidelberg New York Dordrecht London

Printed on acid-free paper

Springer Japan KK is part of Springer Science+Business Media (www.springer.com)

Preface

The role of agriculture, especially the production of food crops, in enhancing food security in the developing world is well recognized. Therefore, research and development agencies are generating and promoting innovations in these crops, many of which are focused on improved seeds of farmers' preferred crop varieties. The improved seeds have captured the attention of these agencies, especially after the Green Revolution of the 1960s as the performance of modern varieties remained better to local landraces with reference to wider adaptability, responses to inputs such as fertilizers and pesticides, and resistance to pests and diseases. The impact of these varieties in bringing significant improvement in food availability in the world is already acknowledged. Efforts are still going on in the promotion of these varieties in farming communities of the developing world where food insecurity, poverty, and climate stress are important issues. Despite this, there is a concern that modern varieties could not reach to small and marginal farmers of the developing countries due to physical, institutional, and socioeconomic constraints. On the other hand, landraces evolved from on-farm innovations could not receive public supports for their conservation and utilization on a wider scale. Studies have shown that landraces supply over 80 % of the total seed requirements in rural areas of the developing world, especially in open-pollinated varieties of food crops including rice and wheat. This does not mean that there is no value of modern varieties in small farmers' conditions. These farmers can be benefitted by modern varieties if they are adapted in low-input conditions. These issues necessitated developing and strengthening institutional mechanisms at local levels that supply crop varieties evolved from modern plant breeding activities and farmers' innovations simultaneously.

The process of searching for alternative institutional mechanisms to deliver seed in rural areas was started from the 1990s, when multinational companies and government-owned corporations could not be cost-effective in supplying seeds of open-pollinated crop varieties at local levels. These mechanisms are the establishment of community seed banks, and empowerment of farmers' social organizations or their associations in seed production and marketing. However, the latter has been more effective to increase farmers' access to diversified varietal choice and is becoming popular in the name of "community seed production (CSP)" systems. This system is also called an integrated seed system because it catalyzes innovations in formal and informal seed systems, and enhances their impacts in food security and livelihoods of the people. In CSP, farmers produce seed at household levels and sell into the markets through their organizations. Development projects, mainly those led by non-governmental organizations, are engaged in promoting CSP although the involvement of government organizations in this initiative has also begun to increase in recent years. Studies have shown that, except in a few cases, development projects have failed to establish successful CSP and their efforts have been criticized as artificial. Experiences and innovations made by development projects in CSP are also poorly documented, and the available information is mainly limited to projects' internal reports with limited or no empirical basis. Moreover, there are variations in modalities in implementing CSP across the crops, cropping systems, and the continents, which have resulted in variations in understanding the dynamics of CSP. Therefore, sustainability of CSP has remained a big concern among these stakeholders, and they explain it in their own way based on their professional backgrounds and experiences. Authors internalized this concern while going through the existing literature and presenting some of the research findings at various occasions of international conferences held in Asia (Japan, China, Vietnam, Nepal), Europe (Croatia), and Africa (Ethiopia). One of the recent conferences was the Expert Consultation Workshop on Community Seed Production, organized by the Food and Agricultural Organization (FAO), International Crops Research Institute for Semi-arid Tropics (ICRISAT), International Center for Agricultural Research in Dry Areas (ICARDA), International Agricultural Research Center for Tropical Agriculture (CIAT), and Catholic Relief Services (CRS), in Addis Ababa, Ethiopia, 9–11 December 2013. The objective of the conference was to share issues, experiences, and innovations made by CSP experts in Asia, Africa, and Latin America. The participants presented their papers, intensively discussed the key issues, and finally came to the consensus that, (1) CSP has been adopted in different crops, farming systems, and continents in diverse modalities, (2) Sustainability is the most important concern in CSP, and economic dimensions alone could not capture the essence of sustainability, and (3) It is necessary to measure the performances of the CSP system using empirical evidence to make this system understandable in academic and policy circles.

This book attempts to analyze sustainability of the CSP system under the rice–wheat farming system, which is the dominant food production system of the Indo-Gangetic Plain, and is being spread across different parts of the world, from the micro-economic perspective, by putting seed producers and consumers in the context. In doing so, it has utilized the empirical data associated with rice and wheat seed production where necessary. How seed producers realize benefits and how that benefit continues in the future is the main thrust of this book. Seed producers' performance in resource management, marketing, and governance cover their current benefits. Similarly, soil conservation and risk-management strategies provide the basis for their future benefits.

There are 11 chapters in this book. The initial three chapters are introductory. Chapter 1 discusses the historical background, spatial dispersion, and production challenges of the rice–wheat farming system. The concept of CSP, its evolutionary history, and implementation modalities are discussed in Chap. 2. Chapter 3 presents a framework to analyze the sustainability of CSP, which provides a foundation for understanding the rest of the chapters that are based on empirical evidences.

Economic dimensions of CSP are discussed from Chaps. 4, 5, 6, and 7. Chapter 4 analyzes households' behavior in buying seed from the market. Results show that households adopt diversified varieties, and a neighboring-farmer is the most important source of improved seeds. Households' behavior in buying seed is mainly explained by irrigation facilities, membership in groups/cooperatives, seed prices, operational land, and education of the household heads. Chapter 5 measures technical efficiency of farmers in utilizing the most commonly used inputs: seed, labor, fertilizer, and land in rice and wheat seed production. Findings indicate the existence of a wide variation in efficiencies among the seed producers, and this variation is explained by the household head's education, land quality, and experience in seed production. Profitability of the seed producer is measured in terms of profit efficiency in Chap. 6, which indicates that male-headed households are more efficient than female-headed ones; and those with irrigation facilities, having better-quality houses and adopted crop-diversification strategies are more profit efficient than their counterparts.

Technical efficiency and profit efficiency capture the essence of economic benefits to be realized by seed producers at the production stage, but could not capture their potential benefits at the marketing stage. Chapter 7 analyzes farmers' behavior in selling seed in the market, and it recognizes the significant roles of family labor, irrigation facilities, operational land, training, share collection, and seed price in households' seed-selling behavior.

It is quite difficult to precisely estimate the continuity of the current level of benefits realized by seed producers in the future due to various uncertainties. However, soil health and risk-management strategies could address this issue. Chapter 8 analyzes farmers' behavior in adopting different soil conservation practices, and it shows that larger operating households prefer zero-tillage practice whereas those having irrigation facilities adopt green manure crops. Chapter 9 measures profit risk in CSP and assesses households' behavior in risk-management strategies. It is clear from this study that households adopt crop diversification and group saving as risk-management strategies. Households with younger household heads and those associated with the Brahmin caste are more likely to adopt group saving practices. Similarly, households with less-educated household heads and smaller operational land prefer crop diversification practice. However, those better off in physical assets, such as irrigation facilities; and aware of socio-economic benefits from seed production adopt both of these practices.

Moreover, seed producers interact with public (research and extension) and private (traders) agencies through their organizations (seed producer organizations – SPOs) to access production inputs such as source seed, and to sell seeds in markets. The governance (participation, business plan development, incentive system, and linkage) of SPOs determines the level of support seed producers receive from these agencies. Chapter 10 measures organizational governance with reference to members' participation, business plan development, incentive systems and linkages; and make its linkage with household-level benefits. Results indicate the positive impact of governance indicators on household-level economic indicators, i.e., profit efficiency and proportion of seed sold by seed producers in the markets. It provides a possibility that the facilitation of organizational governance could be one of the entry points to enhance economic benefit to seed producers. This approach of facilitating CSP is much more important for capacity building of seed producers, which are in their early developmental phase.

Chapter 11 summarizes the key lessons learned from the analytical chapters on various aspects including, dynamic leadership, entrepreneurship, participatory crop improvement, soil health and group saving. It also discusses how the sustainability issues of CSP could be institutionalized in programs and policies of actors involved in the promotion of the CSP system. It also argues that institutionalization is a continuous process, and it enhances adaptive and innovation capacity and therefore provides a basis for the continuity of the system. However, to catalyze the institutionalization process, the mechanisms for the continuous engagement of major actors needs to be developed.

The empirical evidence presented in this book was collected from seed producers and consumers associated with the rice–wheat farming system in the Tarai region of Nepal. We extend our great appreciation to these farmers. Anonymous reviewers evaluated earlier versions of the chapters of this book, which were presented as articles in various journals. We thank them for their comments and suggestions, which have added values to the academic merit of this book. Thanks also go to members of Maharjan's research laboratory, Graduate School for International Development and Cooperation, Hiroshima University, for their contribution in brainstorming ideas and choosing appropriate analytical tools for data analysis. We also recognize the Global Environmental Education Leadership Program of Hiroshima University, Japan, and the Forum for Rural Welfare and Agricultural Reform for Development (FORWARD Nepal), for financial support during the field study.

We believe this book will be useful for students, researchers, policy makers, and donor agencies working with CSP in the developing world to broaden their understanding of CSP in general, and its sustainability in the particular. Attempts have been made to present information in a simple way, minimizing formatting errors. However, all limitations of the book are the responsibility of the authors. We always welcome comments and suggestions from readers for its improvement.

Hiroshima, Japan Narayan Prasad Khanal
Hiroshima, Japan Keshav Lall Maharjan

Contents

About the Authors

Narayan Prasad Khanal, Ph.D., was born in January 8, 1976, in a rural family residing in Kerunga Village Development Committee, Ward No. 1, Arghakhanchi district, Nepal. He is the youngest son of Mr. Churamani Khanal and Mrs. Gun Kumari Khanal. From his childhood, he used to support his parents in carrying out crop husbandry practices. This motivated him to continue his higher education majoring in agricultural science. He completed his Intermediate (1994), Bachelor (2000) and Master (2002) degrees in agriculture from Institute of Agriculture and Animal Science (IAAS), Tribhuvan University, Nepal. He also completed Ph. D. in Agriculture Economics (2013) from Graduate School for International Development and Cooperation (IDEC), Hiroshima University, Japan, with financial assistance of Japanese government. He has already gathered about 10 years' working experience at national and international levels including IAAS as a 'Research Associate', National Grain Legumes Research Program as a 'Legume Expert', Forum for Rural Welfare and Agricultural Reform for Development (FORWARD Nepal) as a 'Senior Program Manager', and IDEC as 'Assistant Professor'. Currently, he is working with International Maize and Wheat Research Program (CIMMYT), South Asia Regional Office, Kathmandu, Nepal, as an 'Agriculture Economist / Business Development Expert'.

Mr. Khanal has designed, implemented and evaluated over a dozen research and development projects in Nepal, in partnership with national and international organizations. These projects were related to poverty reduction, sustainable rural livelihoods, crop intensification, climate smart agricultural technologies, agricultural biodiversity, and participatory approaches. Some of the lessons from these projects have also been institutionalized in the Nepalese government systems, one of them to enumerate includes release of mungbean varieties (Kalyan and Prateeksha) by combining data generated by government and non-government organizations in Nepal in 2006. He has also shared his experience and lessons in national and international forums held in Asia, Africa, Europe and America. Mr. Khanal has published 11 peer reviewed research articles, 5 proceeding papers, 3 booklets, and a couple of policy documents. He is also actively engaged in agricultural and environmental related networks, and committed to contribute pro-poor rural innovations.

Keshav Lall Maharjan (Dr. of Agriculture in Agricultural Economics, Kyoto University, Japan) is currently a Professor at the Graduate School for International Development and Cooperation, Hiroshima University, Japan, where he has been teaching, conducting research and chairing various steering and decision-making committees since its foundation in 1994. He gives lectures for graduate students on subjects including Rural Economics, South Asian Studies, International Development and Cooperation Studies. He conducts weekly seminars at the graduate school that address pertinent issues in Agricultural Economics, Rural Development, Sustainable Development, Cultural Dynamics, Climate Change and Rural Livelihood Strategies in developing countries. He also offers support for graduate students writing their Master's theses and doctoral dissertations on the related topics of Development Sciences, Educational Development and Cultural and Regional Studies, which include issues concerning natural resource management, food security, poverty dynamics, local governance, rural society and community dynamics. In doing so he considers agriculture and rural regions as not only the source of cheap labor, cheap food and cheaper intermediate inputs, and subordinate to urbans and centers as marginal sector and peripheral region but also as a dignified way of life for people who are guardians of nature and are more conscious about the earth, humans and their interaction, so as to sustain this culture and civilization for generations hereafter. Rural regions are such places that make these things happen. Hence, fieldwork to grasp the diverse realities of rural regions location specifically before generalizing the research is given importance in his research, lectures and in educating the graduate students in terms of their research, writing journal articles and dissertations. Some 25 students have received their Ph. D. from Hiroshima University under his guidance.

In order to disseminate research findings, consolidate ideas and concepts, and share knowledge with other professionals, he regularly participates in local, national and international seminars and conferences organized by academic societies, research institutions, various organizations and like-minded individuals, including agricultural economists, ruralogists, sociologists, environmentalists, anthropologists, educationalists, policy makers, development practitioners, farmers, social activists, local leaders and opinion shapers.

Some of his earlier books in English include *Communities and Livelihood Strategies in Developing Countries*. Tokyo: Springer Japan, 2014, *Climate Change, Agriculture and Rural Livelihoods in Developing Countries*, Tokyo: Springer

Japan, 2013, *Understanding Maoist Conflict in Nepal: Initiatives of Civil Societies on Social Capital Development for Peacebuilding in Hills*, Germany: Lambert Academic Publishing, 2013, *Peasantry in Nepal: A Study on Subsistence Farmers and Their Activities Pertaining to Food Security*, Hiroshima: Research Center for Regional Geography, Hiroshima University, 2003 and *Impacts of Irrigation and Drainage Schemes on Rural Economic Activities in Bangladesh*, Hiroshima: Research Center for Regional Geography, Hiroshima University, 1997. He has also contributed chapters to publications including *Microfinance, Risk-taking Behaviour and Rural Livelihood*, New Delhi: Springer India, 2014, *Geography of governance: Dynamics for local development*. International Geography Union Commission on Geography of Governance, Slovakia, 2013, *Climate Change: Asian Perspective*, Jaipur: Rawat Publication, 2012; *Globalization and Cultural Practices in Mountain Areas: Dynamics and Implication*. Sikkim: INDUS, 2012, *Public Policy and Local Development - opportunities and constraints*, International Geographical Union Commission on Geography and Public Policy, 2008; *Political and Social Transformation in North India and Nepal*, New Delhi: Manohar Publishers, 2007; *Contentious Politics and Democratization in Nepal*. New Delhi: Sage Publications, 2007, *Small-Scale Livelihoods and Natural Resource Management in Marginal Areas of Monsoon Asia*, Dehra Dun: Bishen Singh Mahendra Pal Singh, 2006; *New Challenges Facing Asian Agriculture under Globalization*. Selangor: Malaysian Agricultural Economics Association, 2005; *Translating Development: The Case of Nepal*, New Delhi: Social Science Press, 2003; and *Sustainable Agriculture, Poverty and Food Security*, Jaipur: Rawat Publications, 2002. He contributes to various related academic journals and has more than 125 blind reviewed articles to his credit. He has also produced numerous books and journal articles in Japanese.

Acronyms and Abbreviations

AE	Allocative efficiency
AIC	Agriculture Input Corporation
amsl	Above mean sea level
BLM	Binary logistic model
CBOs	Community-based organizations
CH_4	Methane
CIAT	International Center for Tropical Agriculture
CO_2	Carbon dioxide
COB	Client-oriented breeding
CRS	Catholic Relief Service
CSP	Community seed production
DADOs	District Agriculture Development Offices
DAP	Diammonium phosphate
DFID	Department for International Development
FAO	Food and Agriculture Organizations
FAOSTAT	FAO statistics
FORWARD	Forum for Rural Welfare and Agricultural Reform for Development
FYM	Farm yard manure
GOs	Government organizations
HHH	Household head
IAAS	Institute of Agriculture and Animal Science
ICARDA	International Center for Agricultural Research in Dry Areas
ICRISAT	International Crops Research Institute for the Semi-Arid Tropics
IFPRI	International Food Policy Research Institute
IGPs	Indo-Gangetic plains
K	Potassium
LFU	Labor force unit
LSU	Livestock standard unit

MoAC	Ministry of Agriculture and Cooperatives
MNL	Multinomial logit
MVP	Multivariate probit
N	Nitrogen
NARC	Nepal Agriculture Research Council
N_2O	Nitrous oxide
NGO	Non-government organization
NRs	Nepali rupees
OLS	Ordinary least square
OM	Organic matter
OP	Open-pollinated
P	Phosphorus
PCA	Principal component analysis
PE	Profit efficiency
PPB	Participatory plant breeding
PVS	Participatory variety selection
RMPs	Risk management practices
SCP	Soil conservation practice
SD	Standard deviation
SFPM	Stochastic frontier production model
SML	Simulated maximum likelihood
SPIN	Special Program in Nepal
SPOs	Seed producers' organizations
STATA	Data analysis and statistical software
SRI	System of Rice Intensification
SQCC	Seed Quality Control Center
TE	Technical efficiency
USAID	United States Agency for International Development
UVP	Univariate probit
VDC	Village Development Committee
VDI	Variety diversification index
VIF	Variance inflation factor
ZT	Zero-tillage

Chapter 1
Rice–Wheat Farming at a Glance

Abstract The rice-wheat (R-W) farming system supplies 45 % of the digestible energy and 30 % of the total protein requirements of the world. This system is popular in the Indo-Gangetic plain and is being extended to different parts of the world to address increasing food demands. This system also holds enormous potential for intensification and diversification of croplands in tropical and subtropical regions. Although there was a substantial increment in rice and wheat production in the world after the Green Revolution, the yield of these crops has been stagnant in recent years. Major reasons for this are declining soil organic matter, increasing production costs, erratic weather patterns, and poor access to quality seeds of the varieties preferred by farmers in rural areas.

Keywords Production challenges • Sustainability • Diversification • Cereal yield

1.1 Introduction

Rice (*Oryza sativa* L.) and wheat (*Triticum aestivum* L.) are the world's two most important cereal crops, contributing 45 % of the digestible energy and 30 % of the total protein in the human diet (Evans 1993). In addition, straw from these crops is used as feed for livestock, as fuel for household energy, and as thatching materials. Rice is the predominant crop of the tropics, but wheat predominates in the temperate region. Historical evidence shows that people have grown wheat after rice in a sequence on the same land, which is also known as the rice-wheat farming system, since the Tang Dynasty period (617–907 AD) in China (Lianzheng and Yixian 1994), and since 1872 AD in Uttar Pradesh of India (Timsina and Cornor 2001). However, with the development of photo-insensitive and short-duration rice and wheat varieties, especially after the Green Revolution in the 1960s, the rice-wheat (R-W) farming system became more popular in both low-altitude and highland areas.

The importance of the rice-wheat farming system has been recognized more in recent years to address the issue of food insecurity and poverty in developing countries, especially in the Asia and the Pacific regions. These crops contribute 56 %, 60 %, 62 %, 94 %, and 63 % of the total calorie intake in China, India, Pakistan, Bangladesh, and Nepal, respectively (Timsina and Cornor 2001).

N.P. Khanal, K.L. Maharjan, *Community Seed Production Sustainability in Rice–Wheat Farming*, DOI 10.1007/978-4-431-55474-5_1

Moreover, this region is the world's most populous place, accommodating 60 % of the global population, two thirds of the world's poor, and 60 % of the world's undernourished people (IFPRI 2012). A total of 563 million poor people, which represent 13.9 % of the total population, are found in this region. This situation demands development and promotion of mechanisms that contribute to sustainable intensification of the rice-wheat farming system. This chapter discusses the coverage of the rice-wheat farming system, characteristics of the system, major production challenges, and potential measures to address these problems.

1.2 Coverage of Rice–Wheat Farming

Basically, the rice-wheat farming system has been spread from tropical to warm temperate regions of South and East Asia, and these areas are characterized by cool and dry winters and hot and wet summers (Timsina and Cornor 2001). This system is concentrated in the Indo-Gangetic Plains (IGPs) into the Himalayan foothills, covering a vast area from Pakistan's Swat Valley to the north to India's Maharashtra State in the south, and from mountainous Hindu Kush Afghanistan in the west to the Brahmaputra foot plains of Bangladesh in the east. The IGP is composed of Indus (areas of Pakistan, and parts of Haryana and Punjab in India), the Gangetic Plains (Nepal, Bangladesh, and some parts of India such as Uttar Pradesh and Bihar). About 85 % of the R-W cropping system of the IGPs is in South Asia (Evans 1993). Woodhead et al. (1993) found that the major adopters of this system are in India (10 Mha) and China (13 Mha) (Fig. 1.1). In China, this system is adopted along the Yangtse River basin (Jiangsu, Zhejiang, Hibei, Guizhou, Yunnan, Sichuan, and Anhui Provinces) and along the Yellow River basin. The other countries of the IGPs have a smaller area under this system, such as Pakistan (2.2 million ha), Nepal (0.6 million ha). and Bangladesh (0.5 million ha). Moreover, The practice of growing wheat after rice is also common in Bhutan, Iran, Japan, Korea, Philippines, Myanmar, Thailand, Indonesia, Vietnam, Sri Lanka, Mexico, and Egypt, although on a small scale (Timsina and Cornor 2001).

1.3 Characteristics of Rice–Wheat Farming

1.3.1 Diversification

Generally, the rice-wheat farming system involves growing two crops (rice and wheat) in sequence in a year, but there is substantial diversity in cropping patterns under this system. In the western part of the upper IGP, only two crops, monsoon rice (aromatic varieties are also popular, under the brand name of Basmati) and spring wheat, are grown in a year, and there is not sufficient time for the integration

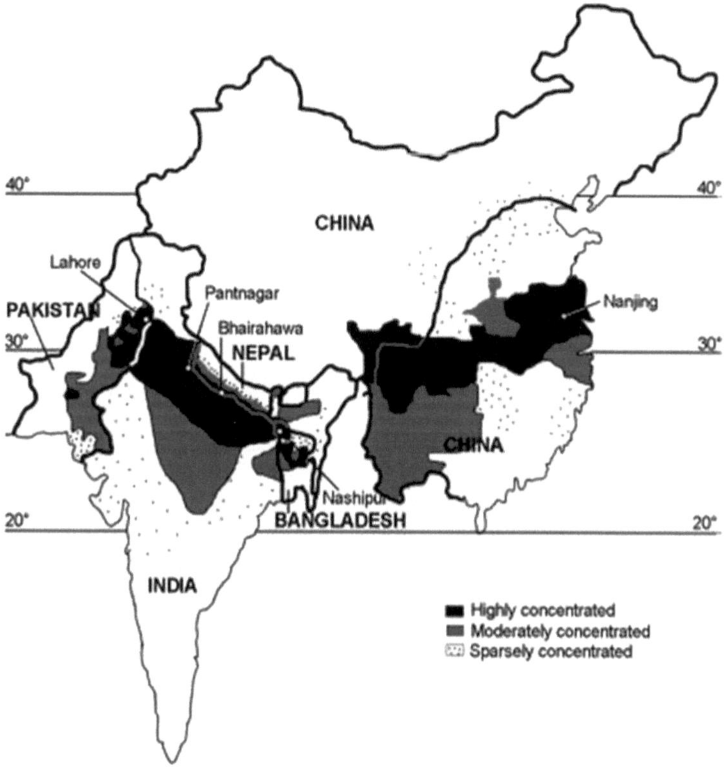

Fig. 1.1 Distribution of rice-wheat farming system in Indo-Gangetic Plain (From Timsina and Cornor 2001)

of a third crop. In this domain, rice is transplanted in May/June and harvested in September/November. Then, wheat is grown in October/November and harvested in March. In the eastern part of the upper IGP, and in the middle and lower IGPs, rice (mainly in those places where indica varieties are popular) is transplanted from June/July and harvested in October/November, and wheat is grown from November/early December to March (Timsina and Cornor 2001; Ladha et al. 2003). After harvesting of wheat, the land remains fallow for about 60–120 days, and farmers, especially smallholders, grow different crops to diversify their livelihood portfolio. The diversification strategy also contributes to averse risk among small farmers (Khanal et al. 2012). Crops being integrated in rice-wheat-fallow include maize (*Zea mays*), mungbean (*Vigna radiata*), cowpea (*Vigna unguiculata*), dhaincha (*Sesbania* spp.), aus/spring season rice, jute (*Chorchorius* spp.), sweet potato (*Ipomoea batatas*), and potato (*Solanum tuberosum*).

Small farmers also grow wheat in mixed cropping patterns with lentil (*Lens culinaris*) or chickpea (*Cicer arietinum*), as these patterns increase cropping intensity and minimize the risk of crop loss. In some irrigated areas, mustard and vegetable crops have replaced wheat. There is also variation in the ways of sowing wheat/winter crops after rice harvest. In locations having sufficient soil moisture and less infestation of weeds, winter crop seed is broadcast inside the rice fields about 7–14 days before the rice harvest (Ladha et al. 2003). In that case, farmers do not normally apply chemical fertilizers and organic manure, assuming that the residual nutrients remaining after the rice harvest would be sufficient for winter crops (Khanal and Maharjan 2010).

1.3.2 Changes in Soil Property

The normal practice of rice transplanting involves puddling the soil to form a saturated root zone for the ready establishment of transplanted seedlings. The puddling scheme in rice is generally considered advantageous because it enhances resource use efficiency, yield stability, and productivity, primarily by retaining water and nutrients and suppressing weeds (Ladha et al. 2003). However, non-rice crops grown after the rice harvest, such as wheat, show poor performance: one reason is low germination/crop establishment rates resulting from poor contact of seed with soil in cloddy fields formed in the hard soil pans by dispersion of soil aggregates and compression of subsoils (Lianzheng and Yixian 1994). Crops grown in such low-permeable soils might be susceptible to drought because of their poorly developed root system. This soil physical property is also linked with soil chemistry, the combination of which also determines the basis for the availability of nutrients to plants and the emission of greenhouse gases from the R-W cropping system. For example, anaerobic decomposition of organic matter in the waterlogged field produces methane (CH_4), which escapes to the atmosphere by diffusion through the rice plant during its growing season. The major pathways of CH_4 production in waterlogged soils are the reduction of CO_2 with H_2, with fatty acids or alcohols as hydrogen donor, and the transmethylation of acetic acid or methanol by methane-producing bacteria (Dormaar et al. 1988).

1.4 Major Production Challenges

The rice-wheat farming system is facing two major challenges, the first being related to the economic concern of low productivity and stagnation of crop yield. The second challenge is the potential negative impact of the rice-wheat farming system on the natural resource base and global environment. This section discusses these two challenges.

1.4.1 Economic Concern

The yield of rice and wheat has been substantially increased in the past 50 years (1961–2010) in the world (Figs. 1.2, 1.3). For instance, average rice yield in the world in 1961 was 1,869 kg ha^{-1} but it was more than doubled (4,336 kg ha^{-1}) in 2010. The progress in wheat yield is similar to that of rice. Wheat yield was 1,008 kg ha^{-1} in 2001 but had increased almost threefold by 2010 (3,003 kg ha^{-1}). However, the yield growth rate of these crops started declining in the 1990s. The average growth rate of rice yield was 2.71 %, 1.6 %, 2.47 %, 1 %, and 0.9 % per year in the 1960s, 1970s, 1980s, 1990s, and 2000s, respectively. Similarly, the average wheat yield growth rate remained at 3.7 %, 1.4 %, 3.6 %, 1.1 %, and 1.2 % per year in the 1960s, 1970s, 1980s, 1990s, and 2000s, respectively (FAOSTAT 2014). Moreover, the increase in yield is different across countries. Among the countries compared, China has a higher yield growth rate in both rice and wheat as compared to the world average and other countries.

The yield of these crops has also been fluctuating across the years.

Some of the reasons for declining yield growth are increasing costs of cultivation, declining soil fertility, lack of good-quality seed, and scarcity of water (Timsina and Cornor 2001; Tripathi et al. 2006; Khanal and Maharjan 2010). Decline in soil fertility is mainly related to decreasing soil organic carbon and deficiency of micronutrients (Zn, Ca, Fe) in soils, as supported by results of long-term soil fertility trials carried out in research stations and diagnostic tests at farmers' fields (Ladha et al. 2003).

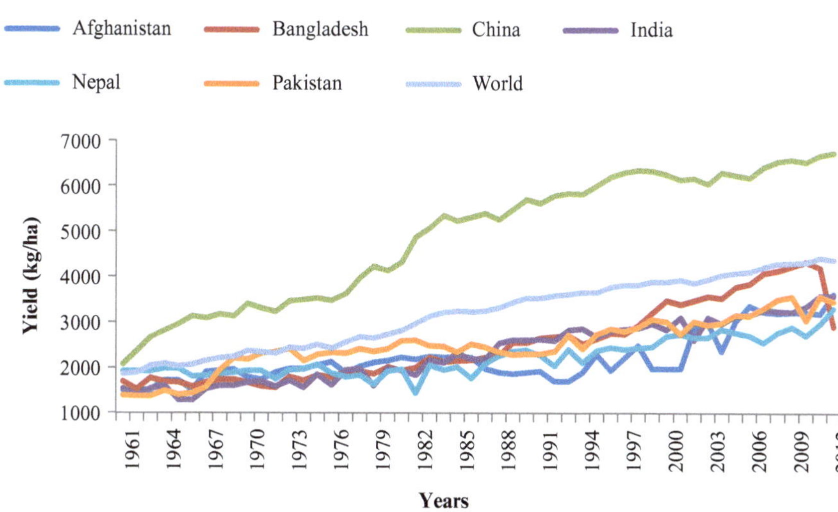

Fig. 1.2 Rice yield trend in different countries (Raw data from FAOSTAT 2014)

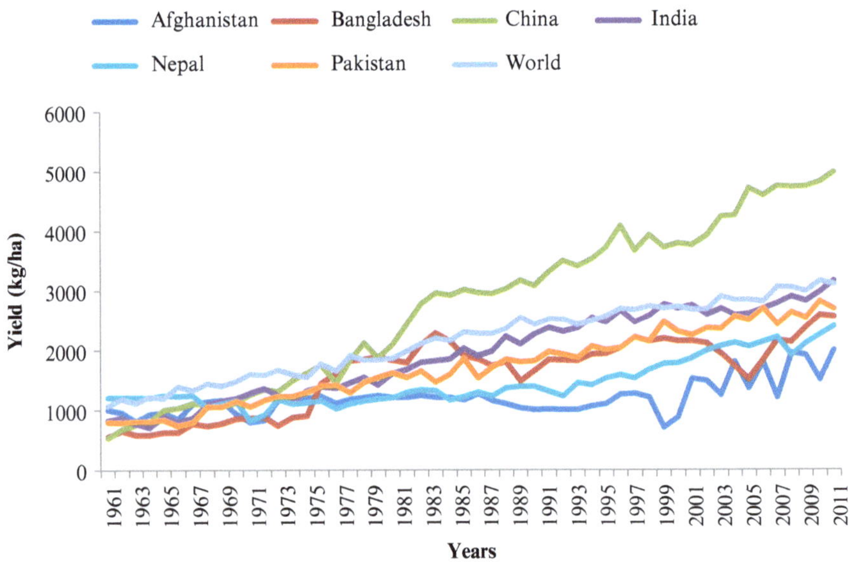

Fig. 1.3 Wheat yield trend in different countries (Raw data from FAOSTAT 2014)

1.4.2 Environmental Concerns

The rice-wheat farming system interacts with the environment from two perspectives. First, most of the farmers produce rice and wheat in a rain-fed environment; therefore, any deviations in rainfall trends will have negative impacts on crop yield. Droughts are the most common climate-induced disaster in rice and wheat production in developing countries. These disasters are the result of acute water shortage, causing severe and sometimes catastrophic economic and social consequences. Grace et al. (2003) found India, Pakistan, Bangladesh, Sri Lanka, and Nepal to be vulnerable to drought. It is estimated that drought in Bangladesh in the 1990s caused a reduction in rice production of 3.5 million tons. Similarly, drought caused substantial reduction in rice and wheat production in Nepal during 2007 and 2008 (WFP 2010). However, drought is not the only climate-related disaster affecting rice and wheat production: other climatic disasters such as rise in sea level, hot winds, floods, landslides, and hailstorms also affect these crops.

The rice-wheat farming system is considered as an intensive system, and efforts are underway to increase its cropping intensity through agronomic measures, which demands more energy because of greater resource consumption (Ladha et al. 2003; Regmi et al. 2009), increasing greenhouse gas emission from the rice-wheat farming system. For example, nitrogenous fertilizers applied in crop fields are responsible for emitting nitrous oxide (N_2O) from this system. The nitrogenous fertilizer preparation process also emits greenhouse gases; that is, every kilogram of nitrogenous fertilizer emits 1.8 kg CO_2 as by-product during the processing period (Grace et al. 2003). CO_2 is also emitted by decomposition of organic matter in the soil, and this is more common in warm and wet soil (Denison 1996). The loss of soil organic matter will have detrimental effects on soil fertility and soil structural stability. Similarly, tillage operations catalyze greenhouse gas emissions from the R-W cropping system. On average, every liter of diesel burned by tillage machinery results in the emission of 2.6 kg CO_2 to the atmosphere. In general, 400 kg CO_2 is generated from each hectare (ha) of land assuming 150 l diesel consumption. In general, higher availability of CO_2 could increase crop yield, but this is unlikely to happen in wheat because of increased temperature, which decreases crop duration and induces a sterility problem. Similarly, in rice the increased temperature might affect the reproductive phase; for example, spikelet sterility might be caused when the temperature exceeds 32 °C. As a result, the yield of this crop might decrease at a rate of 5 % with each increment of 1 °C above 32 °C (Grace et al. 2003).

Methane is another greenhouse gas emitted from the rice-wheat farming system, especially from rice fields. In addition, methane is also emitted through burning of crop residues, and it is estimated that methane emitted from residue burning constitutes about 20 % of the total CH_4 emission from the rice-wheat farming system).

1.5 Practices for Harmonizing Environmental Issues and Crop Productivity

One of the key challenges in the rice-wheat farming system is how to harmonize economic and environmental concerns. The yield of food crops is low and should be increased to feed the growing population on one hand, and on the other, the yield increment initiatives should not have negative impacts on natural resources. To address this issue, agricultural scientists are working to develop appropriate agronomic practices such as zero tillage (Fig. 1.4), a system of rice intensification (SRI) (Fig. 1.5), legume integration (Fig. 1.6), and integrated pest management. The fundamental concept of these practices is how to minimize water and energy consumption, how to build up soil organic matter, and how to integrate crops in lands remaining fallow after wheat harvest. Organic matter has an important role in harmonizing economic and environmental goals because it provides nutrients for crop production and improves soil qualities such as water-holding capacity, soil aeration, soil structure, and soil structural stability. The exact practices used to address these issues might vary across location and types of production systems. In general, practices that reduce energy consumption are zero tillage (Tripathi et al. 2006) and the system of rice intensification (SRI). The SRI practice also improves water use efficiency (Upreti 2008) and builds up soil organic carbon. Although these practices also build up soil organic matter, green manure and composting are common practices that increase soil organic matter (Denison 1996). The detail of how these practices contribute to soil health and agricultural

Fig. 1.4 Wheat production in zero tillage (From authors' photo bank)

Fig. 1.5 Farmers transplanting single seedlings using rice transplanter (From authors' photo bank)

Fig. 1.6 Maize and mungbean intercropping in rice-wheat farming (From authors' photo bank)

productivity is given in recent literature (Timsina and Cornor 2001; Ladha et al. 2003; Gupta and Sayre 2007; Shah et al. 2011; Khanal et al. 2012).

Moreover, improved management of chemical fertilizers could also minimize greenhouse gas emission from the rice-wheat farming system in a high-potential production system. These practices include the use of slow-release nitrogenous fertilizers that increase nitrogen use efficiency, and intermittent irrigation and drainage will further reduce CH_4 emission from rice. The use of calcium nitrate or urea instead of ammonium sulfate and its deep placement instead of surface application can increase its efficiency and plant uptake, and therefore reduces N_2O emission (Lal et al. 1998). Growing leguminous crops after wheat harvest has also been found effective in reducing the emission of nitrous oxide (Pandey et al. 2008), and on the other hand this practice sequesters carbon and improves soil fertility and crop productivity (Subedi 1992; Chen et al. 1997; Gupta and Sayre 2007).

1.6 Seed Management in Rice–Wheat Farming

Seed is the determinant of success in agricultural crops including rice and wheat, and its value is determined by its quality. Seed quality is judged from its potential for crop yield and market value. These two values are determined by the genetic, physical, and sanitary qualities of seed (Almekinders et al. 1994). Genetic quality is determined by growth cycle, uniformity, and special characteristics such as disease resistance, pest resistance, drought tolerance, or tolerance of micronutrient deficiency. Its value is reflected in crop yield and market price as well.

The valuation of seed quality parameters is also influenced by type of seed system. In a formal system, narrow ranges of crop varieties are promoted, and seed quality is judged through strict rules and regulation of the government. However, in an informal seed system, which supplies seeds to the majority of small farmers in the developing countries in open-pollinated food crop varieties, quality parameters are flexible (Tripp 2001). Rather, availability and accessibility of seeds are the concern (Khanal and Maharjan 2010). Farmers intend to maximize genetic variation in the planting material (seed) so that they can minimize the risk of crop failure (Almekinders et al. 1994). One of these concerns associated with the formal seed system is whether the selected crop varieties fit into the existing cropping patterns of smallholder farmers; agricultural scientists evaluate a narrow range of crop varieties in their management without considering constraints that hinder their adoption by small farmers. However, farmers evaluate the value of seeds not only by the potential benefits generated from one crop but those from the cropping system in a year. For example, in irrigated domains farmers might grow maize or mungbean or vegetables in the land remaining fallow after wheat harvest. In this case, earliness might be the more important determinant than crop yield in selecting rice and wheat varieties because it allows farmers time for planting of spring season crops. Also, the same farmers might have land with different soil types and moisture regimes, which demand different varieties of rice and wheat to fit into the existing cropping systems. Similarly, land size matters in choosing types of crop varieties. For instance, small farmers intend to choose more stable crop varieties whereas larger farmers' interest would be maximizing the crop yield by combining high-yielding crop varieties with other inputs including fertilizer and pesticides.

Accessibility is another concern for seed in rural areas. Private companies could not be cost effective while handling small quantities of seed of specific crop varieties in niches. The local level seed management scheme demands being able to address farmers' local specific requirements and to supply them a wide range of crop varieties at affordable prices.

The maintenance of genetic quality in rice and wheat seed production, which are self-pollinated crops, is not as difficult as in cross-pollinated crops, because an isolation distance of 3–5 m between two varieties is sufficient to minimize the risk of outcrossing and mechanical mixture (Randhawa 1983).

1.7 Summary

The rice-wheat farming system is popular in the Indo-Gangetic plain, and it is spreading in other parts of the world to address human food demands. This system is also becoming more diversified in warm humid areas, especially in the subsistence farming system. In the past 50 years, the yield of rice and wheat has increased substantially, but its growth rate is even smaller than the population growth rate in recent years. Some of the reasons for this are poor soil fertility and poor access to improved seeds for farmers' preferred crop varieties in rural areas. Improvement in these concerns demands farmers' access to diversified stress-tolerant crop varieties, both those developed by agricultural research and those evolved through farmers' innovations. In the next chapters we discuss community seed production and its sustainability.

References

Almekinders C, Louwaars NP, Bruijin GH (1994) Local seed systems and their importance for an improved seed supply in developing countries. Euphytica 78:207–216

Chen GX, Huang GH, Huang B, Yu KW, Wu J, Xu H (1997) Nitrous oxide and methane emissions from soil–plant systems: nutrient cycling. Agroecosystems 49:41–45

Denison RF (1996) Organic matters. Retrieved 5 Aug 2011, from http://donlotter.net/lotter_organicag.pdf

Dormaar JF, Lindwall CW, Kozub GC (1988) Effectiveness of manure and chemical fertilizers in restoring productivity of artificially eroded dark brown chermozemic soil under dry land conditions. Can J Soil Sci 68:669–679, Retrieved August 2, 2011, from http://donlotter.net/lotter_ajaa_article.pdf

Evans LT (1993) Crop evolution, adaptation and yield. Cambridge University Press, Cambridge, 500pp

FAOSTAT (2014) Crop information 2011. Retrieved 11 Feb 2013, from http://faostat.fao.org/site/567/DesktopDefault.aspx?PageID=567#ancor

Grace PR, Jain MC, Harrington W (2003) Environmental concerns in rice-wheat systems. In: Proceedings of the international workshop on development of action program for farm-level impact in R-W system of the Indo- Gangetic Plains. 25–27 Sep 2000, Rice-wheat consortium paper series 14, Rice-Wheat Consortium, New Delhi, pp 99–111

Gupta R, Sayre K (2007) Conservation agriculture in South Asia. J Agric Sci 145(3):207–214

IFPRI (2012) Global hunger index: challenge of hunger ensuring sustainable food security under land, water and energy stress. Retrieved 10 Jan 2013, from http://www.ifpri.org/sites/default/files/publications/ghi12.pdf

Khanal NP, Maharjan KL (2010) Sustainability of community-based seed production enterprises in Nepal: institutional issues. Nepal Agric Res J 10:33–40. Retrieved February 15, 2014, from http://pustakalaya.org/eserv.php?pid=Pustakalaya:3772&dsID=NARC2010_NepalAgricultureResearchJournalVol10.pdf

Khanal NP, Maharjan KL, Dangol D (2012) Soil conservation practices for sustainability of rice-wheat system in Nepal: a review. J Int Dev Coop 18(4):11–20. http://ir.lib.hiroshima-u.ac.jp/metadb/up/kiyo/AN10482914/JIDC_18-4_11.pdf

Ladha JK, Hill JE, Duxbury JM, Gupta RK, Buresh RJ (2003) Improving the productivity and sustainability of rice-wheat systems: issues and impacts. American Society of Agronomy, Crop Science Society of America and Soil Science Society of America, Madison

Lal R, Kimble J, Follett R, Cole C (1998) The potential of U.S. cropland to sequester carbon and mitigate the greenhouse effect. Ann Arbor Press, Chelsea

Lianzheng W, Yixian G (1994) Rice-wheat systems and their development in China. In: Parada RS, Woodhead T, Singh RB (eds) Sustainability of rice-wheat production system in Asia. FAO, Bangkok, pp 160–171

Pandey KR, Shah SC, Becker M (2008) Management of native soil nitrogen for reducing nitrous oxide emission and higher rice production. J Agric Environ 9:22–28

Randhawa AS (1983) Seed production in wheat. Indian Farm 1:17–18

Regmi A, Tripathi G, Giri S, Bhatta M, Sherchan DP, Karki KB, Kumar V, Ladha, JK (2009) Improving food security through integrated crop and resource management in the rice-wheat system in Nepal. Retrieved September 25, 2011, from http://irri.org/knowledge/publications/irri-books/integrated-crop-and-resource-management-in-the-rice-wheat-system-of-south-asia

Shah P, Dahal KR, Shah SK, Dangol DR (2011) Effect of tillage, mulch and time of nitrogen application on yield of wheat (*Triticum aestivum* L.). Agric Dev J 8:9–19

Subedi KD (1992) Search for indigenous green manuring species in order to sustain soil fertility (in Nepal) Prabidhi Sangalo. Lumle Agricultural Research Centre

Timsina J, Cornor DJ (2001) Productivity and management of rice-wheat cropping systems: issues and challenges. Field Crop Res 69:93–132

Tripathi J, Adhikari C, Lauren JG, Duxbury JM, Hobbs PR (2006) Assessment of farmer adoption of surface seeded wheat in the Nepal tarai. Rice-Wheat Consortium Paper Series 19. Retrieved 10 Jul 2011, from http://gradeducation.lifesciences.cornell.edu/faculty/individual800

Tripp R (2001) Provision of seed for agricultural development. Retrieved 22 Apr 2011, from http://mcknight.ccrp.cornell.edu/program_docs/litlists/Rach_Biblio_Seed%20Syst_18may06.pdf

Upreti R (2008) System of rice intensification (SRI) experience of Nepal. In: A paper presented in third national symposium on SRI, organized by SRI India at Tamil Nadu University, December 1–3, India. Retrieved September 20, 2011, from http://sri-india.110mb.com/documents/3rd_symposium_ppts/Nepal.pdf

Woodhead T, Huke R, Huke E (1993) Rice-wheat atlas of Pakistan. IRRI/CIMMYT/PARC, Los Banos

World Food Program (2010) The cost of coping; a collision of crises and the impact of sustained food security deterioration in Nepal. United Nations World Food Programs, Nepal Office. Retrieved August 1, 2011, from reliefweb.int/./1295211F858E4D878525762B005E5451-Full_ Report.pdf

Chapter 2
Fundamentals of Community Seed Production

Abstract Community seed production is a local-level seed management system owned by farmers. This system is also called an integrated seed system as it combines formal and informal seed systems to enhance their impacts in farming communities. There are various modalities in implementing community seed production, but the capacity building of a farmers' social organization is quite common. In this model, farmers produce seed at the household level and sell it in the market through their organizations. Typical activities associated with this model include seed system analysis, group formation/selection, enterprise management training, technical training, seed production demonstration, seed quality assurance, seed marketing systems, and institutionalization of the lessons.

Keywords Integrated seed system • Capacity building • Seed producers and consumers

2.1 Introduction

2.1.1 Concept and Definition

The term 'community' refers to people residing within a certain geographic boundary or those having common interests (White 1984). In a community, three things are important to understand: 'Who are the stakeholders?' 'What are their common agendas for engagement?' and 'What are the mechanisms for their engagement?' In the relationship of this concept to community seed production (CSP), these people are the farmers who are residing in one geographic area and are 'engaged in seed production' of food crops. The definition of CSP varies in the literature, but the commonality across them is integration of formal and informal seed systems (Bishaw and van Gastel 2008; Khanal and Maharjan 2013; Sperling et al. 2013). However, we define CSP in this book as "local-level seed management system owned by farmers." The term local highlights production and consumption of seed at the local level. It also implies the access of seed in a cost-effective way at local levels as a consequence of lower production and transportation costs. The phrase "owned by farmers" recognizes ownership and the leading role of farmers in handling seed production activities. However, this concept does not underestimate

N.P. Khanal, K.L. Maharjan, *Community Seed Production Sustainability in Rice–Wheat Farming*, DOI 10.1007/978-4-431-55474-5_2

the roles of other participants, but it highlights that farmers should be in the leadership for managing CSP, and the supporting actors should be treated as facilitators. In CSP, farmers organized in groups or cooperatives start seed production activities in open-pollinated (OP) seeds of food crops such as rice and wheat, although there are other modalities of seed production as well (refer to Sect. 2.3 for detail). The organizations are the traditional social organizations formed in the communities to enhance their members' socioeconomic empowerment, which are consistently called seed producer organizations (SPOs) in this book.

2.1.2 Functioning of Community Seed Production

The CSP is also known as an integrated seed system (Sperling et al. 2013), considering its potentials to combine formal and informal seed systems to enhance their impacts in seed security and thereby food security. As mentioned already, there is quite a variation in CSP implementation modalities, but these variations can be generalized as a function of actors and their actions. A typical feature of this phenomenon is depicted in Fig. 2.1, where actors are divided in accordance to their actions, that is, input supply, production, and consumption. Actors included in input supply are nongovernment organizations (NGOs), government organizations (GOs), and private agencies. These participants provide training, source seed, and seed testing facilities. Seed-producing farmers are classified under production. Using these inputs together with their internal resources of land, labor, and capital, they produce improved truthfully labeled seed (from foundation or breeder seed) at the household level, but sell it in the market through SPOs, because organizations hold higher bargaining power in market chains and could increase economy of scale in product marketing as compared to individual dealings (Hunington 1968). However, the size, form, and nature of interactions of SPOs with facilitators change over time in accordance with the growth stages of an CSP (Fig. 2.1). For instance, in the first stage, GOs and NGOs simply raise awareness in farming communities about the importance of seed and its production techniques. They also motivate private sectors to engage in the seed supply system through developing incentive systems. Once SPOs enter the second stage, after accumulating some knowledge on seed quality and institutional management, GOs and NGOs change their support strategies to SPOs, that is, from input distribution to capacity building and networking. In the third stage, NGO support is withdrawn, providing linkage with GOs and private agencies such as media and banks, among others (FAO 2010). The essence of this approach is to catalyze innovations in formal and informal systems and to enhance their impacts in the food security of people. Not much is clear about the development pathway for CSP beyond stage three, but it might develop new entrepreneurs in seed production and marketing as Morris et al. (1998) explained relative to the evolutionary pathway of maize seed.

Farmers (rice grain producers), NGOs/GOs, and traders represent the major actors involved in seed consumption as they buy seed from farmers. However,

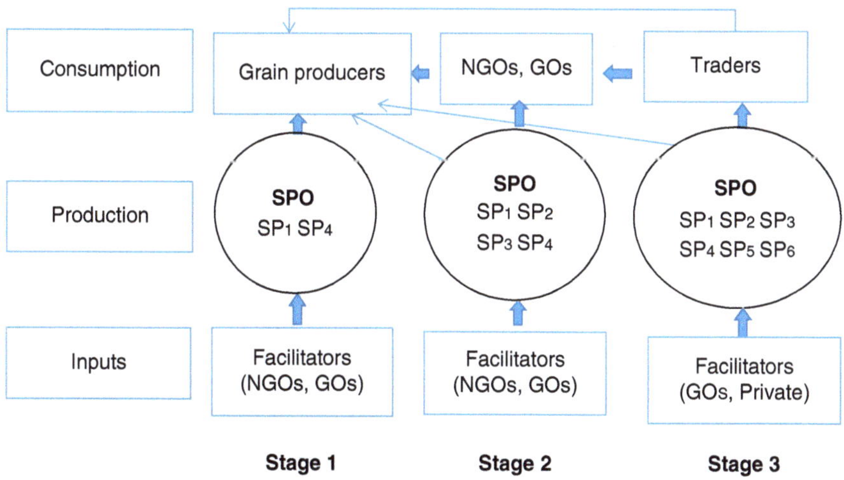

Fig. 2.1 Actors and their roles in community seed production. *SPO* seed producer organization, *GO* government organization, *NGO* nongovernment organization (From authors' own sketch)

the latter two participants are mainly the facilitators; they do not consume seed themselves, but disseminate it to new locations to increase seed demand or to address their developmental goals. At the first stage of CSP, seed producers consume most of the produced seed, and only a small portion is exchanged with neighboring farmers and their relatives (Almekinders and Louwaars 1999). However, in the second stage, most of the seed is consumed by NGOs and GOs, and a minor part goes to the local farmers. When SPOs mature and increase business volume, local traders become interested in seed marketing (Witcombe et al. 2010).

2.1.3 *Benefits from Community Seed Production*

Growing crops for grain production, and keeping part of the produce as seed at the household level or bartering it with friends, relatives, and neighbors, is the traditional practice that has been adopted by farmers since the dawn of agriculture around 10,000 years ago (Tripp 2001). However, it is believed that CSP provides economic, social, and environmental benefits to seed producers and seed consumers (Srinivas et al. 2010; Khanal and Maharjan 2013). These benefits are summarized next at seed producer and seed consumer levels, although grain (food) consumers, traders, and other related agencies also benefit from CSP because of the availability of diverse crops/food choices at local levels.

Fig. 2.2 A farmer happily explaining the benefits of growing seed; farmers' criteria for seed quality judgment (From authors' photo bank)

Seed Producers

(a) Seed producers use quality source seed and improved management technologies in seed production, which increases the yield and value of crop produce. Thus, even though farmers could not sell their produce as seed in the market, especially in the early phase of seed industry development, seed producers will benefit from increased crop yield at lower costs. Food security is improved through increased availability of food grains at the household level at lower costs (Fig. 2.2)

(b) Being a high-volume, low-value commodity, seed can be collected by traders from farmers during the seed sowing period and distributed in a short time (1 or 2 months for a crop) as this substantially reduces storage costs. Instead, farmers could develop seed-processing and seed storage facilities at the local level, which will contribute to a flourishing local economy through increased employment and business opportunities at local levels.

(c) The CSP offers farmers a good platform to share their experience, knowledge, and culture. So, it promotes social cohesion and harmony in the community

(d) While engaging in CSP, farmers' linkage with facilitators and other service providers will be improved. Seed producers could utilize this linkage to develop SPO seed-processing facilities through a public–private partnership approach.
(e) Farmers can improve their entrepreneurship skills and confidence in seed production and marketing while engaging in CSP, which might lead to cultivation of new entrepreneurs in the seed supply chain.
(f) Farmers can address their issues, such as biodiversity, discrimination, and subsidy policies, more effectively while going through their organizations.

Seed Consumers

Seed consumers are also the farmers who buy seed from the market. In the early stage of CSP, farmers residing nearby seed producers represent the main seed consumers. However, the seed consumers' boundary widens in the later stages of CSP development, especially when traders are involved in seed marketing. Seed consumers will benefit from CSP because of the easy availability of a diverse choice of seeds/varieties at local levels at a reasonable price. This varietal diversity accommodates both modern and farmers' varieties, and it provides a mechanism to realize food security in the region through seed security (McGuire and Sperling 2011). Farmers' varieties are more resilient to biotic and abiotic stresses (McGuire and Sperling 2013) and are more valuable from the social and cultural perspective in rural areas.

A typical rural setting is characterized by a diversity of people with respect to livelihoods, assets, and capabilities, and it justifies the necessity for maintaining varietal diversity. Larger operating households intend to maximize their return on investment from crop production by adopting modern varieties in combination with other production inputs. On the other hand, the interests of small farmers would be in adopting farmers' varieties that meet their households' demand (crop yield, straw yield, suitability in the existing cropping pattern, and so on) and help them minimize production risks from various factors such as climate, disease, and pests (Tripp 2001).

Box 2.1: Seed and Its Types

In the botanical sense, seed is the matured ovule or the means of dispersal of plant populations in space and time. It is also called the most efficient vehicle of agricultural technologies, and therefore most of the innovations in this sector are based on genetic manipulations. Moreover, seed quality determining its value is measured in terms of genetic, physical, sanitary, and adaptability characteristics (Randhawa 1983).

Seeds are normally classified into three groups based on the breeding method adopted in their production process: these are open-pollinated (OP),

(continued)

Box 2.1 (continued)
hybrid, and genetically modified seeds, whether they are related to self- or cross-pollinated crops. The OP seeds are produced from natural, random, open pollination by wind, birds, and insects, resulting in plants that are naturally varied. Farmers/researchers carefully select OP seed varieties, which may be self- or cross-pollinated, with respect to their beneficial traits such as tolerance to biotic and abiotic stresses, earliness, and so on. Then, the selected seeds are multiplied, considering appropriate time/space isolation to maintain their genetic purity. It is expected that seeds produced from OP seeds will have characteristics/traits the same as those of their parents.

Crossing two different but related plants develops hybrid seed, producing new desirable traits that cannot be produced by the inbreeding process of the same plants. While using hybrid seed, farmers need to replace seeds every year, because if farmers recycle the produce from hybrid seed, undesirable traits might appear in the grain produced in subsequent generations.

Seeds developed by the genetic engineering process (combining genes from different organisms in the laboratory without the sexual process of breeding) are called genetically modified seeds. For example, genetically modified seeds contain deoxyribonucleic acid (DNA) of a bacterium (*Bacillus thuringiensis*), which kills the European corn borer by punching holes in its gut linings (Tripp 2001). Multinational corporations and large private companies are mainly engaged in the production of hybrid and genetically modified seeds, and resourceful farmers are the major consumers of these seeds. However, the OP seed is common in smallholder farming communities, especially in food crop varieties including rice and wheat. So, we discuss only OP seed production in the CSP system in this book.

Box 2.2: Seed System and Its Classification
A seed system can be broadly defined as the sum of physical, organizational, and institutional components as well as their actions and interaction that determine supply and use of seed in qualitative and quantitative terms (van Amastel et al. 1996). The national seed system constitutes a formal system, informal system, and integrated system. The integrated system is still not well established in the literature and is taken as a means to strengthen formal and informal systems (Bishaw and van Gastel 2008).

Formal Seed System: The formal system is a deliberately constructed and bounded system that involves a chain of activities leading to clear products: certified seed and verified variety (Almekinders and Louwaars 1999). The guiding principles of this system are to maintain varietal purity and identity

(continued)

Box 2.2 (continued)
and to produce seed of optimal physical, physiological, and sanitary quality. Seed production and distribution also follow a limited number of officially recognized seed outlets. This system intends to keep a clear distinction between seeds and grains. In this system, research organizations develop crop varieties using materials stored in a domestic gene bank or imported from foreign countries in partnership with international organizations. These materials are then tested for multiple locations and years with strict supervision of subject matter specialists/agricultural scientists. The best performer lines are released/registered in the name of varieties from the national seed board of the respective countries. The research organizations normally maintain breeder seed (also called source seed) of the released/registered varieties and supply foundation seed (seed multiplied from breeder seed) of these varieties to farmers' groups for further multiplication (Fig. 2.3). Thus, nonreleased/nonregistered varieties do not flow through the formal system (Tripp 2001).

Informal Seed System: The informal seed system, also known as "farmers' seed system" (Almeinders and Louwaars 1999), "traditional seed system" (Lyon and Danquash 1998), and "local seed system" (Almekinders et al. 1994), is the oldest seed system, having been adopted by farmers with the advent of agriculture. In this system, part of the crop harvest is used as seed, or seed is accessed through other informal networks such as friends, neighbors, relatives, and grain markets. The important characteristics of the system are varietal diversity and a not-clear distinction between seed and grain. The same set of functions such as multiplication, selection, storage, and

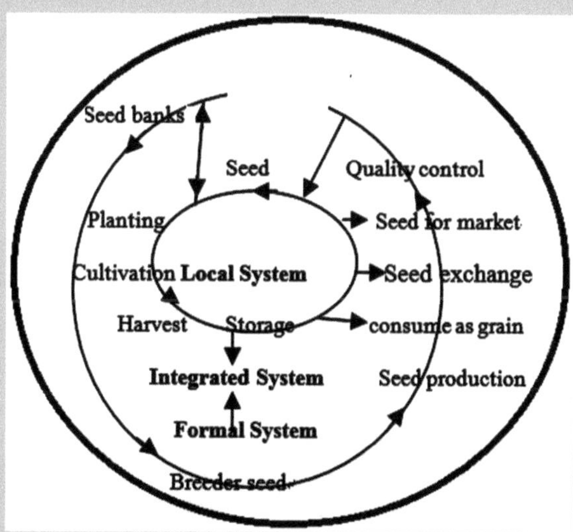

Fig 2.3 Components of national seed system (Adapted from Almekinders and Lowaars 1999)

(continued)

Box 2.2 (continued)
distribution are followed in the informal system as in the formal system. However, farmers consider the informal system as an integral component of the cropping system rather than a bounded discrete activity. This system supplies more than 80 % of seed in food crops, including rice and wheat in developing countries, and is even more prevalent in subsistence farming. Researchers and policy makers intend to enhance the genetic pool in the informal system so that farmers could choose seeds appropriate for their needs (Almekinders et al. 1994).

Integrated Seed System: The integrated seed system indicates coordinated actions between formal and informal seed systems. This term also conveys the interdependence of these systems, with multiple links between the two, with each reaching to the other and changing over time (Sperling et al. 2013). There is a growing interest in integrating formal and informal seed systems at technical and institutional levels to address diversified seed demand at local levels. Adoption of participatory plant breeding is one of the approaches to integrate these systems at the technical level, whereas seed multiplication through local-level institutions integrates these systems at the institutional level. Participatory plant breeding enhances local-level biodiversity by incorporating farmers' seed demand characteristics in the variety development process. Two types of institutions are engaged in seed multiplication: private seed companies, and seed producer groups and cooperatives. The former are focused on hybrid maize and vegetable seeds (more profitable commodities), whereas the latter focus on vegetatively propagated crops and nonhybrid seeds. Thus, the integrated seed system promotes multiple approaches to increase farmers' access to diversified seed (Louwaars and De Boef 2012). Community seed production contributes to the integrated formal and informal system, mainly at the institutional level.

2.2 Evolution of Community Seed Production

The concept of CSP has emerged as a response to the failure of the formal system in disseminating Green Revolution technologies (e.g., modern varieties), especially among smallholder farmers residing in a marginal environment. In the 1960s and 1970s international support aimed to strengthen national seed production programs in developing countries (Cromwell and Wiggins 1993; Almekinders et al. 1994; Mywish et al. 1999; Lyon and Danquash 1998). Some of the programs are the FAO Seed Development and Improvement Program, which was supported in 60 countries during 1972–1984; the World Bank, which supported 13 national seed projects and 100 other seed-related projects during 1975–1985; and USAID, which provided long-term support to public bodies concerned with seed during 1958–1987 (FAO 1998). However, most of the smallholder farmers in the developing world could not benefit from these projects, because government corporations promoted a limited range of crop varieties, generally focusing on large farmers in favorable environments (irrigated area, fertile land), and resource-poor farmers could not adapt (Bishaw and van Gastel 2008). Second, seed production and distribution costs for government corporations remained even higher than that of private sectors because of high transportation and storage costs. Also, these organizations could not supply seed in a timely fashion in rural farm communities because of the complex bureaucratic systems, such as poor accountability of staff in their assigned duties.

There was a policy shift toward disbanding government corporations and encouraging private sector development in cereal seed supply in African countries in the 1980s. However, the commercial seed companies could not be profitable in the production and marketing of OP seeds because of high transaction and storage costs and the unreliable and location-specific demands. In marginal areas, farmers have a tendency to plant multiple varieties to minimize the risk of crop failure from disasters such as drought and diseases. Therefore, commercial companies, even those involved in supplying OP seed initially, turned their business toward hybrids, and supplied open-pollinated seed only if there were bulk demands from development agencies (Almekinders et al. 1994).

From the beginning of the 1990s, NGOs began to be involved in the seed supply system in Asian and African countries. These organizations have the advantage to supply seed in marginal areas because they have a committed staff and can easily win the trust of local farmers as they are engaged in socially valuable but nonprofitable tasks such as poverty alleviation, rural development, and livelihood improvement. In some cases, NGOs are directly involved in seed production, and in other cases they motivate farmers for community seed production through different approaches (David 2004; Witcombe et al. 2010; Srinivas et al. 2010). In the recent years, not only NGOs but also GOs have been engaged in implementing CSP programs. Different modalities of implementing CSP are discussed in the next section.

2.3 Modalities in Community Seed Production

CSP attempts to enhance the impacts of informal and formal seed systems through their coordinated efforts. In this process, formal organizations make innovations in their policies and programs to address the concerns of poor farmers and marginal areas, and informal sectors attempt to enhance farmer capacity and linkage with formal sectors. Basically, there are three modalities in implementing CSP in developing countries: individual approach, mobilization of seed companies, and empowering farmers' social organization.

2.3.1 Individual Approach

The individual approach is quite popular where farmers' groups and cooperatives do not exist or their capacity is very weak. Few progressive and larger operating households having direct contact with project-implementing agencies are selected for seed production. These farmers are given technical training on seed production and supplied with foundation seed. The outcome of this approach is evaluated from the perspective of the number of farmers engaged in seed production and the expected volume of seed produced. Farmers' institutional development and seed marketing approaches are ignored (Mywish et al. 1999). It is hypothesized that seed receivers could serve as resource centers in the community to supply new seeds as well as crop management practices. Sometimes, development agencies also buy seed from these farmers for their dissemination in new locations when agriculture research stations fail to supply seed to these agencies. A case in Tanzania shows that the agricultural research station provided foundation seed to farmers in a rotation basis: those receiving sorghum seed in the first year got groundnut seed in the second year (FAO 2010). The logic behind the crop rotation scheme was to maintain seed quality and to reduce the pressure of soil-borne diseases. The individual approach is cost effective, and easy to implement, but farmers could not learn about seed management activities through their community actions. Also, most of the seed produced under this scheme is more likely to be consumed as grain when development projects fail to buy seed from farmers (FAO 1998). However, farmers become aware of the value of seed for crop production, and its technical dimensions, which might encourage new entrepreneurs in the seed supply system.

2.3.2 Mobilization of Seed Companies

One of the limitations for the poor adoption of modern varieties in marginal areas is lack of appropriate delivery channels. Here, development projects, especially those led by government agencies, mobilize seed companies for seed distribution considering that the relationship between seed producers and seed companies could be enhanced. As a result, some of the farmers (seed producers) could serve as contract farmers for the companies after a few years of implementing this model. The International Crops Research Institute for the Semi-Arid Tropics (ICRISAT) and Seed Company of Zimbabwe piloted this model (Monyo et al. 2004). In that case, ICRISAT provided improved OP crop varieties and offered training and monitoring support in seed production through the company. The company had been producing and marketing hybrid maize seed in the country but was not willing to integrate OP crop varieties of maize and other food crops because of the uncertainty of demand and low profit margin. However, with project facilitation, the company made a contractual arrangement with seed-producing households with the condition that the company would buy produced seed from farmers. The program became successful for the dissemination of groundnut, sorghum, millet, and cowpea varieties in Zimbabwe, Mozambique, and Zambia. However, the success of the program was limited to the project implementation period only, which means mobilizing private companies that also pose a challenge for continuity in seed supply of OP crops.

2.3.3 *Empowerment of Farmers' Social Organization*

This model is common in many developing countries where agricultural extension is focused on a group-based approach (David 2004; Witcombe et al. 2010; Khanal and Maharjan 2014). The common scenario in this case is development projects that motivate existing farmers' social organizations such as groups and cooperatives to act as SPO. These farmers are empowered in technical and managerial aspects of seed production and marketing. This model has the following features.

Seed Producers Take the Organization's Membership The basic idea behind uniting seed growers into SPOs is to enhance their efficiency in seed marketing by improving members' accountability toward their organizations. To be a member of the organization, the farmer pays a membership fee or, in some cases, deposits a share amount (cash deposited by members in their organizations) as per the need in the organization. Moreover, this resource collection strategy would be helpful to increase SPO cash reserves, which can be used to develop a seed-processing facility in the organization (Poudel et al. 2003; Witcombe et al. 2010). Similarly, membership ensures farmers' participation in an organization's decision-making process. Previous studies have shown that organizations collecting a fund from their members have been more successful to attract resources from government agencies and development projects to set up seed-processing facilities (seed storage house, grading machines, etc.) (Witcombe et al. 2010; Khanal and Maharjan 2013).

Farmers' Empowerment Is a Key in the Facilitation of the CSP Program The CSP considers that although the CSP is facilitated in the interface of farmers and service providers, the farmers' role should be the more crucial role; research and development agencies associated with CSP should design their activities in partnership with farmers so that farmers could internalize the ownership of these activities. Some of the activities in which farmers are engaged include selecting appropriate sites for seed production, pests and disease control, rogueing, seed storage and processing, planning, branding, and marketing research. Farmers acquiring these skills are expected to enhance their adaptive and innovation capacity (Witcombe et al. 2010).

Seed Production at a Household Level but Marketing Through Their Organization In CSP, farmers produce seed at a household level but sell their produce through SPOs. It is quite convenient for farmers to handle seed production activities as these activities are similar to grain production except for some technical issues such as seed quality maintenance and use of source seed. However, they sell seed in the market through their SPOs, which have better bargaining power in the market chains, and could minimize transaction costs in seed processing and marketing by increasing business volume.

Entrepreneurship Is Important for the Functioning of CSP As CSP is focused on less-profitable crops and managed by resource-poor farmers in marginal areas, there is great challenge for farmers to generate benefits and allocate benefits generated from this joint venture to the members. This approach demands the selection of potential leaders who are willing to work in a team and have complementary skills and talents for seed management and organizational development. It is believed that leaders with entrepreneurial skills could address issues in the organizations and bring SPOs into a new development phase (Khanal and Maharjan 2014). Some of the factors contributing entrepreneurship skills include previous business experience, education, innovations, and action research. The entrepreneurs generate lessons in various dimensions including minimizing seed marketing costs, promoting cooperation among members, and fostering linkage between SPOs and service providers, developing incentive systems and physical facilities in SPOs.

Community Seed Production Is Dynamic Over Space and Time The CSP has both spatial and temporal dimensions, and the first one is the result of variability in crops considered for production and level of seed industry development. For example rice, wheat, maize, and legumes are the common crops being promoted in CSP in Nepal (Khanal and Maharjan 2010); whereas beans in Central America; and sorghum, pearl millet, chickpea, lentil, sweet potato, and cassava in Africa are the major crops promoted under CSP (Srinivas et al. 2010). Similarly, institutional development in CSP becomes faster in a favorable environment such as better access to communication and physical facilities. Moreover, using the lessons and experience of seed production, the size, form, and rules in CSP keep on changing (Morris et al. 1998).

2.4 Procedure for Implementing Community Seed Production

As already discussed, capacity building of farmers' groups and cooperatives is the most common approach in strengthening CSP in developing countries. However, the exact activities implemented might vary with variations in the level of CSP development and crops considered for seed production.

This section discusses the general procedure being adopted in empowering these organizations in seed production and marketing in Asian and African countries, especially with OP seeds. Seven activities are involved, starting from participatory seed system analysis to institutionalization of CSP (Fig. 2.4).

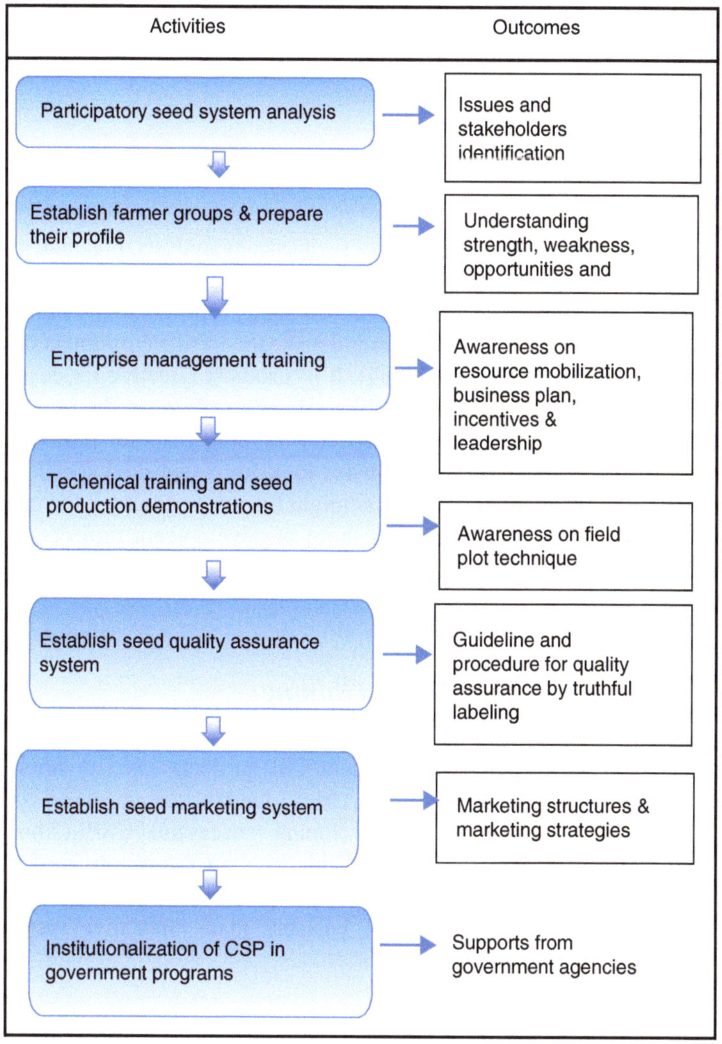

Fig. 2.4 Activities and outcomes of community seed production (Adapted from Bishaw and van Gastel 2008)

2.4.1 Participatory Seed System Analysis

Project staff together with local government professionals interact with local people about major crops and their production systems, problems, and potential solutions. This analysis will also document seed demand and supply dynamics and the challenges for seed sector development. It also maps out potential participants associated with the selected problems and solutions of seed systems. This activity could also arouse interest among the farmers and government agencies about the seed subsector.

2.4.2 Group Selection/Formation

Based on the feedback from participatory seed system analysis, potential groups or cooperatives (social organization) involved in CSP are selected. Justifications for selecting such established social organizations are as follows. Social organizations could easily influence the local situation and make the local environment conducive for project implementation (White 1984). It is also cost effective for development projects while going through these established organizations in the project implementation process because it reduces capacity building costs for projects. It does not mean that development projects always look for established social organizations for project implementation. Projects might form new groups if the interests of these organizations did not match with those of development projects or project locations are not technically appropriate for seed production (Kugbei 2007).

2.4.3 Enterprise Management Training

This concept has been introduced in recent CSP projects when reviews of seed system development projects in Asia and Africa found lack of a business concept as one of the major hindrances for developing successful CSP (Almekinders et al. 1994; FAO 1998). The purpose of this training is to motivate farmers to consider seed production from a business perspective that includes leadership, development and operationalization of business plans, incentive systems, and record keeping (Fig. 2.5).

2.4.4 Technical Training and Seed Production Demonstrations

The several technical considerations in seed production include site selection, seed quality monitoring, seed storage, and seed packaging (Kugbei 2007). Development projects provide technical training to all members of CSP (Fig. 2.6), or they collaborate with government agencies such as agricultural development offices to access resource personnel. However, some members of SPOs receive detailed training on specific issues such as methods of rogueing and local-level seed testing schemes (Witcombe et al. 2010). To handle these issues in an efficient way, SPOs form three subcommittees from their members: a technical subcommittee, a

Fig 2.5 Facilitation of group members for business plan development (From authors' photo bank

marketing subcommittee, and a finance subcommittee. Members of the marketing subcommittee are empowered in estimating seed demands and in designing strategies for seed marketing. Similarly, the finance subcommittee members take training on record keeping, raising group funds, and building organization structures such as a storage house and grading machine.

In addition to providing theoretical training, development projects facilitate establishing seed production demonstrations at farmers' fields. The purpose of these demonstrations is to empower seed producers on field techniques such as maintaining isolation distance between varieties (3 m is required in self-pollinated crops to produce a truthful label seed), and rogueing (Fig. 2.7). These demonstrations are also considered important from the perspective of technology dissemination in rural areas because even those farmers who are not engaged in CSP will have the opportunity to learn integrated crop management practices adopted in seed production.

2.4.5 *Seed Quality Maintenance*

One of the reasons farmers buy seed from markets is its quality, so it is important to capacitate farmers for the adoption of appropriate measures for seed quality maintenance. Truthful labeling is generally recommended for seed quality maintenance in CSP, and in this system, producers assure their quality themselves by putting the SPO's logos and seed quality information such as genetic purity, germination, and physical purity (David 2004). The SPOs collaborate with government-owned seed laboratories to have their seed tested because these organizations do not have their

Fig. 2.6 Farmer training on seed production (From authors' photo bank)

Fig. 2.7 Identification of off-type plants in a seed production plot (From authors' photo bank)

own seed laboratories, and this facility also has not been established in the private sector. The SPOs adopt three strategies to enhance seed quality. The first is the

Fig 2.8 Monitoring of rice field by members of technical subcommittee members (From authors' photo bank)

provision of seed quality management training to their members by mobilizing their technical subcommittee members. Second, SPOs adopt the one-farmer, one-variety approach when distributing source seed to seed growers. Similarly, the technical subcommittee members monitor seed production plots during flowering and harvesting and storage (Fig. 2.8) (Khanal and Maharjan 2013).

At the local level, farmers determine seed maturity level by observing the color of seed and straw, which turns yellow from green during physiological maturity. Seed quality might deteriorate in storage from the time it reaches physiological maturity on parental plants until the farmers plant it. Germination is highest at physiological maturity, and viability then declines inexorably until the seed dies. Deterioration of seed viability cannot be reversed once it has occurred. Good storage cannot improve the quality of poor seed; therefore, only seed with high germination potential and high vigor should be put into storage.

Clements (1987) discussed the problem associated with wheat storage under tropical conditions. Wheat seed is storable for medium to long periods if kept under safe storage conditions. For rice and wheat, high seed moisture (above 11–12 %) is the most damaging, and seed must be kept as dry as possible in storage. The response of wheat seed to high atmospheric humidity (RH) in storage varies with temperature. Clements (1987) reported that at 25 °C and 75 % RH the equilibrium moisture content for wheat is 15 %, and at 90 % RH this may increase to 19.7 %. He also stated that the critical moisture content for wheat that increases the rate of respiration is 14.6 %. In general, stored wheat seed should be kept at moisture content levels below 12 %, and relative humidity should not exceed 50–60 %. Maintaining temperature and moisture is useful to minimize seed rot (e.g., *Aspergillus* spp., *Penicillium* spp.), storage pests [e.g., weevil (*Sitophilus* spp.), khapra beetle (*Trogoderma granarium*), flour beetle (*Tribolium castaneum*)], and seed storage pests (Singh 1985). Farmers traditionally use natural insecticides, such as *neem* (*Azadirachta indica*), biskatali (*Rumex obtusifolius*), and tobacco leaves, as insect repellents for farm-saved seed (Ahmed 1985).

2.4.6 Seed Marketing

Seed marketing is the final step in the CSP; it takes the seed to farmers and gets them to buy it and plant it. Seed marketing is time sensitive and it must reach the farmers at the right time, at the right place, at the right price, in the right amount, and must be of the highest quality. Seed marketing requires convincing farmers that the seed quality is high and ensuring that only high-quality seed is sold; convincing farmers that the seed quality means a benefit to them that is worth the extra cost they pay for the seed. All possible means of promoting the seed must be used, including making the seed available in locations close to the target farmers. The seed must be readily available when the farmers need it, in appropriate-size bags.

Seed producers normally sell seed to neighbors in the early phase of seed industry development, and in some cases development projects buy seed from farmers. However, once the CSP has matured, seed produced by households is collected at SPOs and sold in the market chains. However, it is a challenging task for seed producers to sell seed in the market in the early phase of CSP because of the poor capacity of the SPOs. A case in Nepal shows that government agencies and NGOs could contribute to the promotion of seed produced by SPOs by the provision of subsidies in seed transportation and publicity of seed through demonstrations, interactions, and advertisement (Khanal and Maharjan 2010). In food crops such as rice and wheat, seed is not normally treated with chemical pesticides during storage. If farmers fail to sell crop produce as seed, it is sold in the form of grain but at a lower price (Khanal and Maharjan 2013). This strategy reduces potential loss from seed marketing even if farmers fail to sell seed in the market for various reasons such as unfavorable weather or competition with the traders. Another common strategy adopted by SPOs is diversification in packaging size and seed distribution channels.

2.4.7 Institutionalization

Institutionalization refers to the process of embedding knowledge/concepts within an organization, social system, or society as a whole. This term is frequently used in CSP literature to enhance support for seed producers from government agencies (Khanal and Maharjan 2010), because farmers engaged in CSP represent small-holders residing in marginal environments. Moreover, seed production activities are mainly facilitated by NGOs, whose works are often criticized as being artificial

Fig 2.9 Joint monitoring visit in seed production plot (From authors' photo bank)

because they are mostly donor oriented and lack promotional pathways to scale-up and scale-out lessons (FAO 1998). Studies carried out in CSP in developing countries (Bangladesh, Ecuador, Ethiopia, Zambia, Mali, Mozambique, Nepal, Philippines, and Sudan) have also shown that most of the CSP initiatives collapsed after the project phase-out, and one of the reasons for their collapse is lack of institutionalization (Cromwell and Wiggs 1993). It is believed that those CSPs institutionalized in government programs are expected to receive extension facilities from government agencies (David 2004). Development projects normally adopt two strategies to institutionalize CSP: registration of SPOs in government agencies, and organizing joint monitoring visits (Fig. 2.9). In the latter case, stakeholders including local government officials, media professionals, millers, and traders are invited to observe seed production demonstrations. These stakeholders get an opportunity to observe field activities and discuss how their programs and policies could address farmers' problems in CSP.

2.5 Summary

Community seed production is a local-level seed management system owned by farmers. In this system, farmers produce seed at the household level but sell it to the market through their organization. It provides an institutional mechanism to promote both modern and farmers' varieties at local levels. The CSP has been mainly promoted by development projects, and these projects empower farmers in technical and managerial skills for seed production and marketing. Because seed production initiatives are run by smallholders in marginal areas, CSP should be institutionalized in government programs, which requires understanding of how seed producers and consumers realize incentives in seed production and consumption. This information is necessary to design appropriate policies for enhancing the sustainability of this system.

References

Ahmed SM (1985) Wheat seed production, storage and distribution in Bangladesh. In: Villareal RL, Klatt AR (eds) Wheat for more tropical environments: Proceedings of the international workshop. CIMMYT, Mexico City, pp 291–296

Almekinders C, Louwaars N (1999) Farmers' seed production: new approaches and practices. Retrieved 12 Apr 2011, from www.acss.ws/Upload/XML/Research/99.pdf

Almekinders C, Louwaars NP, Bruijin GH (1994) Local seed system and their importance for an improved seed supply in developing countries. Euphytica 78:207–216

Bishaw Z, van Gastel AJG (2008) ICARDA's seed delivery approach in less favorable areas through village-based seed enterprises: conceptual and organizational issues. J New Seeds 9 (1):68–88

Clements DJ (1987) In: Klatt AR (ed) Wheat seed storage under tropical condition: wheat production constraints in tropical environments. CIMMYT, Mexico City

Cromwell E, Wiggins S (1993) Sowing beyond the state: NGO and seed supply in developing countries. Retrieved 1 Apr 2011, from linkinghub.elsevier.com/retrieve/pii/0305750X9400133J

David S (2004) Farmer seed enterprises: a sustainable approach to seed delivery? J Agric Hum Values 21:387–397

FAO (1998) Developing seed security strategies and programs for food security in developing countries. In: Proceedings of the international workshop on seed security for food security: contribution for the development of seed security strategies in disaster-prone regions. Florence, 30 Nov–1 Dec 1997

FAO (2010). Promoting the growth and development of small seed enterprises for food security crops. Retrieved 10 March 2011, from www.fao.org/docrep/013/i1839e/i1839e00.pdf

Hunington SP (1968) Political order in changing societies. Yale University Press, New Haven

Khanal NP, Maharjan KL (2010) Sustainability of community-based seed production enterprises in Nepal: institutional issues. Nepal Agric Res J 10:33–40

Khanal NP, Maharjan KL (2013) Factors influencing households' behavior in selling rice seed in the market. Agric Food Econ 1–14 http://www.agrifoodecon.com/content/1/1/14

Khanal NP, Maharjan KL (2014) Role of organizational governance on household level economic indicators: evidence from community-based rice seed production of Nepal. J Int Dev Coop 20 (3):13–28

Kugbei S (2007) Seed economics: commercial consideration for enterprise management in developing countries. International center for agricultural research in the dry areas. Scientific Publishers, Jodhpur

Louwaars NP, De Boef WS (2012) Integrated seed sector development in Africa: a conceptual framework for creating coherence between practices, programs, and policies. J Crop Improv 26:39–59

Lyon F, Danquash A (1998) Small-scale seed provision in Ghana: social relations, contracts and institutions for micro-enterprise development. Retrieved 15 Jan 2010, from www.laolink.org/Literature/agrenpaper_84.pdf

McGuire S, Sperling L (2011) The links between food security and seed security: facts and fictions that guide response. Dev Pract 21(4-5):493–508

McGuire S, Sperling L (2013) Making seed systems more resilient to stress. Glob Environ Chang 23:644–653

Monyo ES, Rohrbach DD, Mgonja MA (2004) New partnership to strengthen seed systems in Southern Africa: innovative community/commercial seed supply models. In: Setimela PS, Monyo E, Banziger M (eds) Successful community-based seed production strategies. CIMMYT, Mexico City

Morris ML, Rusike J, Smale M (1998) Maize seed industries: a conceptual framework. In: Morris ML (ed) Maize seed industries in developing countries. Rienner, Boulder

Mywish M, Julie H, Ducan B (1999) Increasing seed system efficiency in Africa: concepts, strategies and issues. Retrieved 15 March 2011, from ideas.repec.org/p/ags/midiwp/54578.html-12k

Poudel DP, Choudhary K, Chowin R, Ghimire H (2003) Seed production and marketing through farmers' groups: case studies. Center for Arid Zone Studies, University of Wales, Cardiff

Randhawa AS (1983) Seed production in wheat. Indian Farm 1:17–18

Singh SB (1985) Production, storage and marketing of wheat seed in India. In: Villareal RL, Klatt AR (eds) Wheat for more tropical environments: Proceedings of the international workshop. CIMMYT, Mexico City

Sperling L, Boettigers S, Barker I (2013) Integrating seed system: planning for scale brief 3. Retrieved 10 Aug 2014, from www.Seedsystem.com

Srinivas T, Bishaw Z, Rizvi J, Niane AA, Manan AR, Amegbeto K (2010) ICARDA's approach in seed delivery: technical performance and sustainability of village-based seed enterprises in Afghanistan. J New Seeds 11(2):138–163

Tripp R (2001) Provision of seed for agricultural development. Retrieved 22 Apr 2011, from mcknight.ccrp.cornell.edu/program_docs/litlists/Rach_Biblio_Seed%20Syst_18may06.pdf

Van Amstel H, Bottema JWT, Sidik M, van Santen SE (eds) (1996) Integrating seed systems for annual food crops. In: Proceedings of a workshop 24–27 Oct 1995. CGPRT, Bogor Indonesia

White LG (1984) Lessons in development project management. University of Maryland, College Park

Witcombe JR, Devkota KP, Joshi KD (2010) Linking community-based seed producers to markets for a sustainable seed supply system. J Exp Agric 46(4):425–437

Chapter 3
A Framework for Analyzing Sustainability of Community Seed Production

Abstract Sustainability issues in community seed production have emerged from economic and environmental routes, but the first approach is more common. Concerns associated with the economic route are marginality, artificiality, and poor capacity. Marginality indicates that farmers involved in community seed production mostly reside in remote areas with poor communication, transportation, and government services. Similarly, development project motivation for farmers without entrepreneurship skills to engage in seed production raises an artificiality issue. The third concern is related to the engagement of poor and smallholder farmers in seed production. The environmental concern is how to prioritize local landraces instead of modern varieties in seed production. This chapter proposes a framework for analyzing the sustainability of community seed production from the perspective of how seed producers realize the benefit and how that benefit is sustained. Household efficiency in seed production and marketing indicates economic benefit, adoption of soil conservation practices shows environmental performance, and organizational governance indicates social performance.

Keywords Economic benefit • Organizational governance • Soil conservation practices • Environmental performance • Social performance

3.1 Introduction

The term 'sustainability' is derived from the Latin word 'sustinere,' meaning to sustain, endure, support, or continue (Onions 1964). It is commonly used in recent years in the context of 'sustainable development' when Brundland's report defined "sustainable development as the development that meets the needs of the present without compromising future generations to meet their own need" (WECD 1987, p. 14). This definition highlights the necessity for designing development interventions in such a way that the current challenges are met but without deterioration in the quality of natural resources that provide the basis of our living on the planet. When adapting this concept in community seed production (CSP), sustainability is the nondeclining benefit of farmers from seed production. This chapter discusses sustainability issues in CSP, summarizes the considerations for analyzing the

N.P. Khanal, K.L. Maharjan, *Community Seed Production Sustainability in Rice–Wheat Farming*, DOI 10.1007/978-4-431-55474-5_3

sustainability of CSP, and presents a framework to analyze the sustainability of CSP in the rice-wheat system.

3.2 Sustainability Issues in Community Seed Production

Sustainability in community seed production (CSP) comes from two routes: economic and environmental. This section discusses major issues related to these routes.

3.2.1 *Economic Route*

The concerns associated with the economic route are marginality, artificiality, and poor government policy. CSP initiatives in developing countries are implemented in marginal areas (rain-fed area, poor soil fertility, remoteness, poor farmers). Concern is raised how the seed producers could be cost effective in these areas with low crop yield and limited communication and infrastructure facilities (Bishaw and van Gastel 2008; Khanal and Maharjan 2010). The second concern is how entities motivated by nongovernmental organizations (NGOs) could manage a seed enterprise as entrepreneurship is neglected in the programs and policies of these organizations (Almekinders et al. 1994). This concern became eminent when reviews showed the poor performance of seed support programs in Asia and Africa (FAO 1998) in strengthening CSP initiatives. It was found that development projects focused on seed distribution, and in many cases these projects bought seed back from farmers, paying a higher price than usual. As a result, once projects stopped buying seed from seed producers, many of these producers dropped out.

The third economic concern is poor government policies in strengthening CSP. As already discussed, seed producers are poor farmers, and they operate seed production activities in marginal areas. It is expected that government programs and policies address the concerns of seed producers. However, there are concerns in government policies, some of which are as follows.

Training Programs Do Not Prioritize Institutional Governance Issues Good governance is one of the success indicators of any joint venture, including CSP, and it is measured by democracy, participation, equity, and empowerment (Gray and Kraenzle 1998). Good governance in the organization enhances cohesion, and accountability among the members, and thereby members' economic benefits from seed production. However, governance issues have been neglected in government agency training programs. The training programs are primarily focused on technical aspects of seed production such as maintaining isolation distance and rogueing. However, CSP faces the challenge of addressing concerns of the poorer members as they might not participate in seed production or may drop out for economic reasons. The SPOs in developing countries have followed a cooperative modality for their organizational setup, and this modality is expected to solve this problem. However, an analysis of 24 cooperatives from developing countries shows that addressing poor members' concerns might hamper an organization's economic performance (Cook 1995). For example, if a one-member-one-vote principle applies in the decision-making process, farmers who supply a large volume of produce have no more say than minor suppliers, and this results in inadequate investments from members in SPOs.

Lack of Mechanism to Provide Follow-Up Support in CSP It is generally recommended that government agencies provide follow-up support to NGO-facilitated SPOs (David 2004). This support includes business development, seed quality maintenance and marketing training, and seed money to develop seed-processing facilities (Crissman and Uquillas 1992; Tripp et al. 1998; David 2004; Bishaw and van Gastel 2008). In a newly established CSP, NGOs organize meetings with government agencies to share their project outputs and status of SPOs, and government agencies make verbal commitments to support SPOs, but these commitments are less likely to be transformed into reality because of limited financial resources and human capital in the organizations (David 2004).

Inefficient Seed Quality Control Mechanism Truthful labeling, a system of seed quality assurance in which producers themselves assure their seed quality, is the most efficient and economical method of seed quality management in CSP (Almekinders et al. 1994). This system is recommended for private seed entrepreneurs including SPOs for seed quality assurance (David 2004; Tripp 2001). Although the truthful labeling system is provisioned in seed policies of developing countries, it is poorly implemented (Lal et al. 2009).

Limited Access to Financial Services Access to finance is a key to successful operation of any business, and CSP cannot be the exception. Because the time for planting and harvesting of crops is seasonal in nature, the SPOs need to collect seed from producers and store it for as long as 7 months before selling in the market. This system requires prompt payment for seed to seed producers, especially the poorer members, at the time of seed collection. If SPOs fail to do so, seed might be sold as grains or farmers could sell their seed to other traders for economic reasons. On the other hand, finding loans for SPOs in rural areas of developing countries is a challenging issue because microfinance institutions are mainly concentrated in city areas (Pradhan 2009; Witcombe et al. 2010). Moreover, these institutions consider it is risky to lend money to SPOs, considering their relatively poor management structure and handling seeds of low-profit crops. As a result, SPOs have to demonstrate collaterals such as land certificates, even if financial institutions agree to provide loans (Pradhan 2009).

Limited Access to Source Seed One of the main reasons farmers buy seeds from markets is to obtain good-quality seeds of their preferred varieties (Almekinders et al. 1994). The demand for seed is generally higher once the variety becomes popularized than it is in the initial stage of popularization. It is therefore important for SPOs to prepare an appropriate mix of popular existing varieties and new varieties. Agricultural research organizations are the main suppliers of source seed to SPOs, and therefore the types of varieties included in source seed production and their quantity influence SPOs. Studies have shown that agricultural research organizations primarily make a foundation seed production plan that is not based on the demands of SPOs but considers the availability of breeder seeds at research stations (Poudel et al. 2003). The breeder seed is primarily composed of varieties tested in a high-input management condition in researchers' controlled conditions. Thus, these varieties cannot address concerns of poor farmers such as adaptation in a low-input condition.

Participatory crop improvement approaches, such as client-oriented breeding and participatory variety selection, have been promoted to address this issue (Witcombe and Virk 1997). These approaches are believed to integrate farmers' criteria in the variety development and selection process, and some varieties have been released through these approaches as well. For example, in Nepal few rice and mungbean varieties developed through the participatory approach have been released, but the source seed supply of these varieties is a challenging issue (Joshi et al. 2012). Similarly, the government of Vietnam does not allow for registration of varieties bred by farmers (Witcombe and Virk 1997). All these issues indicate that it is quite a challenging task for SPOs to maintain an appropriate mix of varieties in a seed production plan. The SPOs could also consider nonreleased varieties in seed production, but producing source seed of these varieties by themselves is again difficult because of their poor technical capacity and resource endowments. Another concern associated with source seed in CSP is how to get a sufficient quantity of the varieties available at agricultural stations. The foundation seed (source seed) produced in these stations is quite a lot less than the requirement. A case in Nepal shows that government organizations supply a negligible quantity of source seed (Table 3.1). Although some private firms are licensed to produce foundation seed, none of them has started supplying foundation seeds of cereal and legume crops because of their poor capacity (SQCC 2013).

In addition to government policy in the provision of facilities in seed production, programs and policies of NGOs also matter. Reviews of the development projects intended to support CSP in Asia and Africa have shown that these projects focus on achieving their development goals of poverty reduction by mobilizing such seed producers' groups/cooperatives in disseminating seeds/agricultural technologies, and less emphasis has been given to their capacity building (entrepreneurial skills). As a result, farmers produced seed but failed to sell their seed in the market as per their plan (FAO 1998).

Table 3.1 Foundation seed supply of major food crops in Nepal

Crop	Area (ha × 1,000)	Annual requirement (*t*)	Quantities of seed certified (*t*)					
			2004	2005	2006	2007	2008	2009
Rice	1,549	19,362	229.1	193.9	232.2	244.5	327.3	253.0
Maize	870	6,090	17.7	12.9	17.1	6.7	10.6	124.8
Wheat	706	21,180	118.4	70.6	92.7	88.9	180.1	43.6
Lentil	265	1,988	0.96	NA	0.2	2.5	1.02	NA

Source: SQCC (2013)

3.2.2 *Environmental Route*

The second sustainability issue in CSP is the environmental aspect. There are three environmental concerns in CSP: (1) how to prioritize local landraces in seed production, (2) how to minimize negative impacts of climate variables in seed quantity and quality, and (3) how to maintain good soil health in the seed production fields.

The concern for prioritizing local landraces in CSP came from farmers' right activists and ecologists, and their argument is that more than 80 % of the seeds of major food crops in developing countries are supplied by farmers' varieties but government support is focused only on modern varieties, and farmers' innovations are neglected, a consequence of the influence of agricultural scientists (plant breeders and agronomists), who are mainly guided in a top-down approach. The ecologists' idea for prioritizing local landraces in CSP came from the perspective that these resources are quite adaptable and serve as cheaper sources of crop variety at a local level.

The second environmental concern is the risk of deteriorating quantity and quality of seed as a result of climatic vagaries. Seed production in developing countries is carried out in open fields and in many cases under rain-fed environments. So, any alteration in climate variables such as rainfall and temperature will have a detrimental effect in farmers' performance in seed production. For example, shifting of the monsoon later than the normal planting season reduces seed crop yield and seed quality. A similar situation occurs in the crop yield of wheat and its seed quality with changing rainfall patterns. Delaying in wheat planting is also associated with increased sterility problem in wheat.

The third environmental issue associated with the rice and wheat (R-W) seed production system is negative concerns of these crops for soil health and the global environment. It has been widely realized that soil organic matter is declining in the R-W system across the production domain (please refer to Chap. 1 for details).

3.3 Framework for Analyzing the Sustainability of Community Seed Production

3.3.1 *Some Considerations in the Analysis*

From an economic perspective, sustainability is nondeclining benefits. However, it is important to clarify whose benefits are considered and how to measure those benefits. When CSP is placed in a system context, it captures the interaction of seed producers and consumers; the former intend to enhance their welfare by maximizing benefits from seed production but the latter's interest is to access quality seeds of appropriate varieties in appropriate locations and costs. Moreover, it is difficult to measure the benefits realized by these factors directly because generating profit from seed production is not the only mission of farmers being engaged in SPOs. Rather, these farmers could realize social (participation) and environmental (biodiversity conservation) benefits using the SPO platform. Similarly, beneficiaries of seed production do not represent people only from the current generation but also from the future generation. How to measure and balance current and future benefits is another challenge because such matters as biological species, organizations, market mechanisms, and market structure continue to change in space and time. However, maintenance of soil health including ecosystem services could be a more logical way to consider future benefits for the people. So, how seed producers and seed consumers realize benefits and how these benefits continue in the future is the major thrust of sustainability analysis (Brinkerhoff and Goldsmith 1990). Producers intend to increase their benefits by maximizing crop yield and selling crop produce at a higher price whereas the intention of the consumer is to access seed at a cheaper price at the proper time and location and in suitable packaging. Again, how to balance economic, social, and environmental benefits realized by farmers is another issue. However, seed production is the economic activity, so social and environmental benefits should be analyzed linking them with economic benefits.

3.3.2 *Economic Performance in Seed Production*

Seed producers realize economic benefit at two stages: seed production and seed marketing. Seed production is the household-level activity, and at this stage households could maximize their benefit by proper allocation of inputs. This benefit could be measured in terms of technical efficiency and allocative efficiency (Farrell 1957). Morris et al. (1998) argued that realization of efficiency gain by a household at the production level could motivate them in the marketing stage. Seed producers could also increase their benefits by increasing the price of their outputs by adopting value addition and increasing economy of scale; this reduces seed producers' transaction costs in seed marketing. Previous studies have shown that demographic (age and education of household, family size), economic (operational land, irrigation facility, fertilizer, soil or land characteristics), and institutional (membership in the organization, access to training) variables are associated with farmers' production efficiency (Rana et al. 2007; Idiong 2007; Piya et al. 2012).

3.3.3 Relationship of Economic Performance with Environmental and Social Issues

As discussed in the previous section, the economic performance of farmers in seed production is determined by their capacity to maximize crop yield with the available resources and by understanding their capacity to increase output price. Environmental and social performance emerges within these two contexts. The environmental concern is mainly related to risk management activities adopted by farmers and soil conservation practices. These practices address the farmers' current benefits as well as the potential to generate benefits in the future. The social issue in CSP is the participation of members in the decision-making process. This participation is mainly concerned in the seed-selling phase that is carried out by seed producers through SPOs. Selling seed in the market involves series of activities starting from demand estimation, processing, storage, price determination, and distribution. So the social issues in this case are how the concerns of poor members are addressed and how members participate in the decision-making process (White 1984).

As discussed, seed production is primarily an economic activity, and therefore the participation issue should be analyzed within the context of that economic activity. There is variation among members in terms of priority setting. Poorer members look at SPOs from the aspect of conformance, and they think that SPOs design policy in favor of the poor members' concerns by provision of credit or timely payment for seed. On the other hand, the better-off members (those having better economic status) look at organization performance from the perspective of their return on investment. Generating benefits from seed selling might be their primary concern, and they might consider that poor members' participation would be beneficial to them as well because it reduces transaction costs. In addition to these internal factors, seed producers need to maintain relationships with service providers such as government agencies and development projects. To manage these internal challenges and to make linkages with external actors, SPOs form an executive committee from their members following democratic principles. The executive body handles these issues by enhancing economic efficiency in selling seeds through designing policies for good governance in the organization. The good governance in SPOs could be understood by analyzing their policies for member participation, incentive system, business plan development, and linkage with service providers (Gray and Kraenzle 1998; David 2004; FAO 2010). The incentive policy makes the members accountable toward their SPOs, and motivates better-off (more resourceful) members to invest in the organization. This plan also explains how to manage the concerns of free riders and horizons and influence cost and volunteer issues in the analytical framework.

3.3.4 Economic Performance in Seed Consumption

In rural areas, there is heterogeneity of farmers in their socioeconomic characteristics such as land size. Normally, larger farmers tend to adopt modern varieties in combination with other agricultural inputs such as chemical fertilizers or pesticides. On the other hand, the small farmers' priority might be to grow crop varieties that need less external input and are more risk averse in nature. Similarly, seed price, characteristics of crop varieties, cropping pattern, and land characteristics affect the behavior of farmers buying seed from the market (Nkonya and Norman 1997; Paudel and Matsuoka 2008). Whether farmers consume seed could be understood by analyzing their behavior in buying seed from the market, assuming that farmers buy seed from the market considering not only one benefit (such as crop yield) but also analyzing the overall benefit they tend to get from the cropping system.

3.3.5 Components of the Framework

Producers Producers represent rice and wheat seed-producing households having membership in SPOs. Because seed production is carried out at a household level, it is the household decision that converts inputs (land, labor, capital, and source seed) into raw seed using technology (knowledge) and achieves economic and environmental performances (Fig. 3.1). The economic performance gained by a household is efficiency, whereas environmental performance (here it is adoption of soil conservation practice and risk aversion strategies) shows the basis for continuing economic benefits for a long time. Households produce raw seed using land, labor, capital, and source seed. After producing raw seed, they supply it to their organizations (SPOs) for marketing. Then, SPOs convert it into processed seed (with value addition and labeling) and sell to the consumers. The performance of SPOs in marketing can be measured in terms of efficiency and governance. The governance of SPOs is mainly related to how the organizations form rules and regulations in line with achieving efficiency in marketing. Good governance in the organization is also needed to manage conflict/risks that emerge from internal and external factors, because it defines the incentive system for members to work for their benefits in a collective way. Governance also affects the flow of information within the organization and provides the basis for implementing a monitoring and evaluation system. Moreover, it guides how democracy is implemented in the organizations. Participation of members in the decision-making process is the major way of applying democratic principles in the organization. However, it is integrated in the

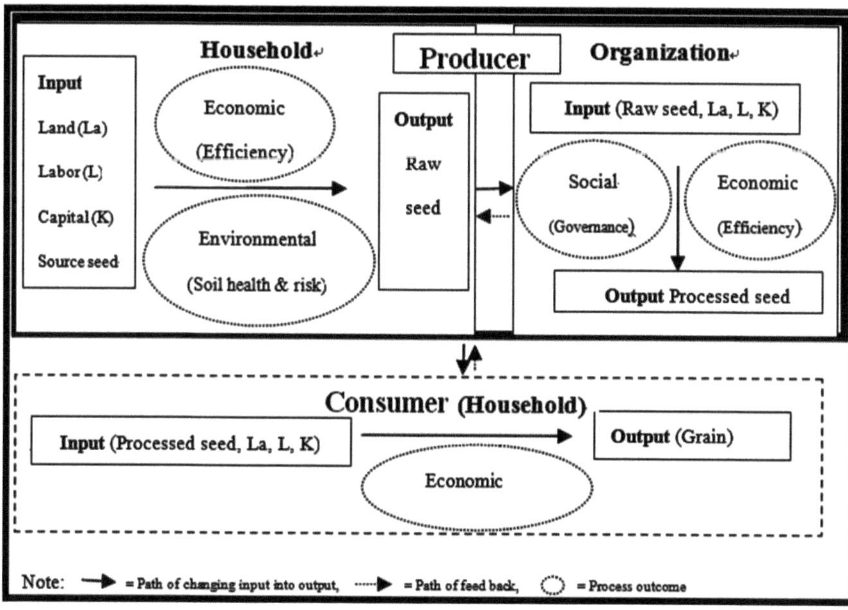

Fig. 3.1 Conceptual framework of the study (Source: Authors' own sketch

governance system as a means to reduce risk against inefficiency of organization in seed marketing.

Consumers Consumers are also rice and wheat farming households residing near the seed producers, but they grow these crops for food and not for seed. These households serve as potential buyers of seed produced by the seed producers. These farmers might get processed seed from SPOs and convert it into rice and wheat grain using resources such as land, labor, and capital. In the process of conversion, if they realize a benefit it is more likely that the consumers would continue buying seed from the seed producers. The benefit might be in the form of crop yield, straw yield, or suitability of varieties in the cropping pattern or market price. However, varietal choice and cheap price could address the consumers' concerns. The adoption of seed/variety is an economic issue, and environmental and social issues are not focused at this level.

3.3.6 Relationship Between Producers and Consumers

Three theories (system theory, contingency theory, and political economy) explain the potential relationship between producers and consumers. The system theory discusses how formal collectivities to informal codes of conduct work in the process of converting source seed into processed seed. The system theory is concerned with the internal process and the relationship between the system and its environment. It thus forces us to think about a wide variety of social, economic, political, technical, and other factors involved in seed production and the marketing process. In other words, it enables us to merge agro-environmental, economic, and managerial aspects in sustainability analysis, because interaction of these elements determines the current and future benefits.

The system theory, however, provides little guidance about how to portray internal system processes or changes in response to externality because the optimal structure or management styles of the production system are contingent on uncertain and exogenous conditions. Contingency theory thus shares with system analysis a concern for environment. The assumption in the theory is that any human aggregation or pattern of behavior has to be seen in relationship to the context of outside forces that threaten or promote its survival and expansion. The contingency theory fills this gap and demonstrates how producers can best attain congruence with the influences of external factors.

Producers can also impart direct influence on consumers. For example, marketing strategies such as seed quality, quantity, location, time of distribution, and publicity could change consumers' behavior in buying seed. The producers could also increase their bargaining power while dealing with service providers such as government agencies, traders, and so on. The households' phenomena to organize in groups/association can be discussed with the help of political economy (Hunington 1968). So, the foregoing framework could be useful to analyze sustainability of CSP because it explains economic, environmental, and social dimensions, putting seed producers and consumers in context.

3.4 Research Design

The analytical chapters of this book (Chaps. 4, 5, 6, 7, 8, 9, and 10) are based on the survey data collected from Nepal during October to November 2011. This section introduces study area and data collection methods adopted in the survey.

3.4.1 Description of Study Area

General Background of the Tarai Region Field survey was carried out in the three districts Siraha, Chitwan, and Kailali, and these districts represent the whole Tarai region of Nepal (Fig. 3.2) as they capture the geophysical, climatic, and socioeconomic variations in the region (Fig. 3.3a, b). This region lies in the southern part of Nepal, bordering with India, and covers the altitudinal range from 70 to 650 m above sea level. It occupies 17 % of the total lands of Nepal but is considered the 'food basket' of the country because more than 60 % of the major food crops including rice and wheat consumed in this country is produced in

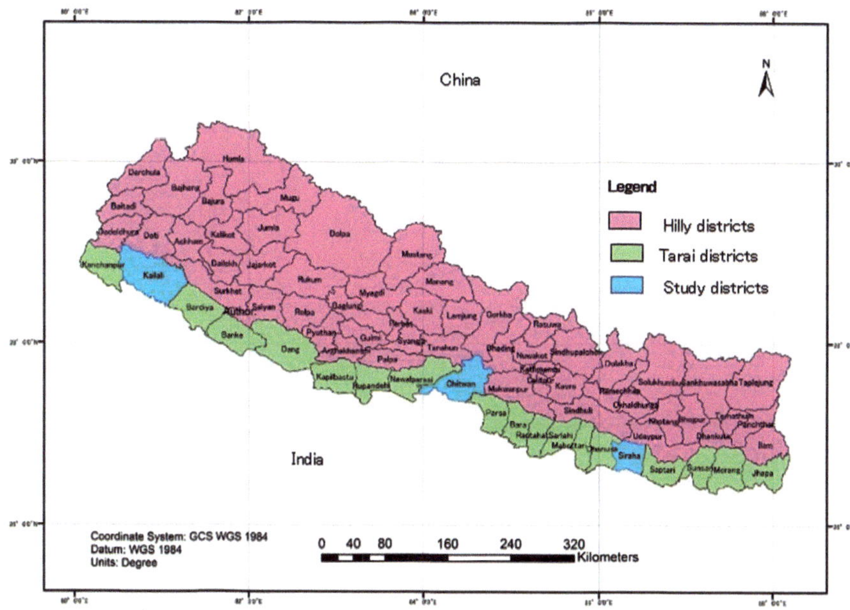

Fig. 3.2 A map of Nepal showing the study districts (Source: Base map from www.ICIMOD.ORG)

the Tarai. Availability of flat land; better irrigation facilities and communication and road facilities; and better soil fertility are some of the reasons that make this region potential for commercialization (Maharjan 2003). Considering this fact, the Nepalese government has emphasized the sustainable intensification of lands in this region through different policies and programs. One of the popular programs to address land intensification is through increased farmer access to diversified crop varieties choice through CSP (Pokhrel 2012) because the formal system supplies less than 10 % of the seed requirement at the national level (Khanal and Maharjan 2010). In this regard, different initiatives were implemented in the country to support CSP from the 1990s. For instance, the 'National Seed Act' and 'National Seed Policy' were formed in 1998 and 2000, respectively. The latter paved the way to involve private agencies in the seed subsector. The District Seed Self Sufficiency Program (DISSPRO) was introduced in 1998 (Poudel et al. 2003). This program envisaged creating three seed centers (two in the Tarai, and one in the mid-hills) across the five development regions. The selected districts were to produce certified first-generation seed from foundation seed (FS) supplied by research stations. Later, the area covered by DISSPRO increased to 63 districts (Chand and Karki 2005).

In addition to the government's led initiatives, several donor-supported projects were also implemented with the objective to help establish the CSP system in the country. Some of these projects focusing on cereal seed promotion in the Tarai region include the Special Program in Nepal (SPIN) project, funded by the Food

a

b

Fig. 3.3 (**a**) Crop production in the study area (Source: Authors' photo bank) (**b**) Livestock production in the study area (Source: Authors' photo bank)

Table 3.2 Characteristics of the study area

Characteristics	Siraha	Chitwan	Kailali
Households adopting agriculture as major source of livelihood (%)	80.5	70.0	79
Land area (km^2)	1,228	2,238	3,235
Cultivated land (%)	60.9	44.3	27.5
Irrigation land (%)	24.8 (21.9)	52 (34)	42.2 (23.3)
Average operational holding per household (ha)	0.721	0.552	0.994
Average family size per household	5.6	4.27	5.26
Annual per capita income (US$)	426	951	583
Average year of schooling (years)	2.82	5.01	3.62
Rice production area (ha)	61,000	29,655	65,500
Rice yield (t ha^{-1})	2.87	3.49	3.1
Wheat production area (ha)	15,725	8,750	34,450
Wheat yield	1.9	3.1	2.07

Source: MoAC (2013), UNDP (2014)

and Agriculture Organization (FAO), which was implemented from 1995 to 1997 in six Tarai districts. Later, the Department for International Development (DFID)-funded projects focusing on participatory crop improvement in cereals and legumes were implemented in Tarai districts from 1997 to 2006. Similarly, another DFID-funded project focused on strengthening CSP was "Participatory Crop Improvement in South Asia," implemented from 2008 to 2012 and covering the entire Tarai districts (Lal et al. 2009; Witcombe et al. 2010; FORWARD 2013).

Characteristics of Survey Area Agriculture is the major livelihood of people in the study area as this sector engages 70 % of its people in agriculture (Table 3.2). Sirha is rather smaller in land size (1,228 km^2) as compared to Chitwan (2,238 km^2) and Kailali (3, 235 km^2) but the proportion of its cultivated land is rather higher than that of the other two districts. Chitwan district is better off than Sirha and Kailali in terms of access to irrigation facilities. Benefits from the irrigation facility will be even less during winter and spring seasons because irrigation canals connected to a stream or river are the major irrigation sources, and water levels in these sources go down during these seasons. In other words, any deviation in the climatic variables such as rainfall patterns would have direct influence in crop production. Farmers of the study districts are smallholders with operational land less than 1 ha per household, and the land size of households from Sirha and Chitwan is even less than the national average landholding (0.8 ha per household) (CBS 2003). Similarly, there is also quite a variation in area and production of rice and wheat across these districts. It is clear from Table 3.2 that area under rice and wheat production is less in Chitwan as compared to the other two districts but the yield level is the highest among these districts.

Moreover, the average yield of rice (3.49 t ha^{-1}) and wheat (3.1 t ha^{-1}) in Chitwan is even greater than the national level of rice (3.3 t ha^{-1}) and wheat (2.42 t ha^{-1}) yields (MoAC 2013). Households make their income from various sources such as food crops, livestock, salaried jobs, and remittance; however, agriculture is the most common source of income across the locations. The income of the people is almost double in Chitwan as compared to the other two districts, which might be the result of commercial orientation in farming from better communication, education, roads, irrigation, and institutional services (Witcombe et al. 2010; Piya et al. 2012; Khanal and Maharjan 2013).

3.5 Sampling Technique

This study used both primary and secondary data. The primary data consist of information collected from household surveys and group discussions. A multi-stage random sampling technique was employed while selecting households for survey. The detailed procedures used for identifying households as seed producers and consumers are discussed next.

3.5.1 Seed Producers

A total of 12 SPOs, 4 in each of the aforementioned three districts, were purposively selected using the list of these organizations registered in District Agriculture Development Offices. From that list, SPOs having at least 2 years' experience in rice or wheat seed production were selected (Table 3.3).

It was found that the number of registered SPOs was quite limited (Chitwan, 10; Kailali, 6; Siraha, 5), and not all the members of these organizations were involved in seed production. Thus, 15 seed-producing households from each of the SPOs were randomly selected for survey. This sampling technique applies to the analytical Chaps. 5, 6, 7, 8, 9, and 10.

3.5.2 Seed Consumers

Seed consumers are also the farmers who grow rice and wheat for grain purposes (for food) and not for seed. These farmers were chosen from nearby villages of the selected SPOs (Fig. 3.4), and it was assumed that they could truly represent the potential buyers of seed produced in CSP. As in seed producers, 15 households were also chosen randomly from one of the selected villages, for a total of 180 households for survey. Moreover, one group discussion in each village was organized to triangulate some of the issues captured in the household survey and to collect village-level information relevant to rice production and consumption (Fig. 3.5).

Table 3.3 Profile of community-based seed producers selected for the study

District	VDC/ municipality	SPO name	Years of establishment	Total members	Involved in seed production	Surveyed households
Kailali	Munuwa[a]	Kisak[d]	2001	58	28	15 (53.7)
	Tikapur[b]	Kisan[c]	1997	26	20	15 (75)
	Masuriya[a]	Sayapatri[b]	2009	20	15	15 (100)
	Chaumala[a]	Kalika[c]	1999	18	15	15 (100)
Subtotal				120	78	60 (80)
Siraha	Padariya[a]	Fulbari	2009	20	19	15 (78.9)
	Gadha[a]	Sagarmatha[c]	2007	25	20	15 (75)
	Gadha[a]	Janadibya[c]	1998	25	23	15 (65.2)
	Siraha[b]	Sampaid[d]	2009	20	15	15 (100)
Subtotal				90	77	60 (77.9)
Chitwan	Patihani[a]	Unnat[d]	2003	98	64	15 (23.43)
	Parwatipur[a]	Shreeram[d]	2003	54	45	15 (33.33)
	Saradanagar[a]	Pragati[d]	2001	74	48	15 (31.25)
	Madhabpur[a]	Bij Bridhi[e]	1998	48	28	15 (53.57)
Subtotal				270	185	60 (32.43)

Figures in parentheses indicate the proportions of the concerned seed production organization (SPO) members

Full names of SPOs: Krisak = Bij Bridhi Krisak Sahakari Sanstha; Kisan = Krisak Bij Bridhi Krishi Sahakari Sanstha; Sayapatri = Sayapatri Biu Utpadak Krishi Samuha; Kalika = Kalika Biu Utpadak Samuha; Janadibya = Janadibya Krishi Sahakari Sanstha; Fulbari = Salhes Fulbari Biu Utpadak Krisak Samuha; Sagarmatha = Sagarmatha Bahuudeshiya Sahakari Santha; Sampaid = Sampaid Biu Utpadan Samuha; Unnat = Unnat Bij Bridhi Krisak Samuha; Shreeram = Shreeram Bij Bridhi Krisak Samuha; Pragati = Pragati Bijbridhi Krisak Samuha; Bij Bridhi = Bij Bridhi Company

[a]*VDC* Village Development Committee

[b]Municipality

[c]Cooperative

[d]Group

[e]Producer company (converted from group in 2006)

3.6 Summary

The sustainability issue of the CSP system has emerged mainly from economic and environmental routes. However, the former is more established, because farmers engaged in CSP represent smallholders who carry out seed production activities in low-value crops, and government services to support CSP are poor. This chapter discusses the major sustainability issues in CSP and provides a framework for analyzing CSP, putting seed producers and seed consumers in the background. How seed producers realize benefits and how those benefits continue in the future is the main thrust in the analysis. The capacity of households in mobilizing resources in seed production and marketing represent the potential economic benefit they realize from seed production. Adoption of soil conservation practices in seed

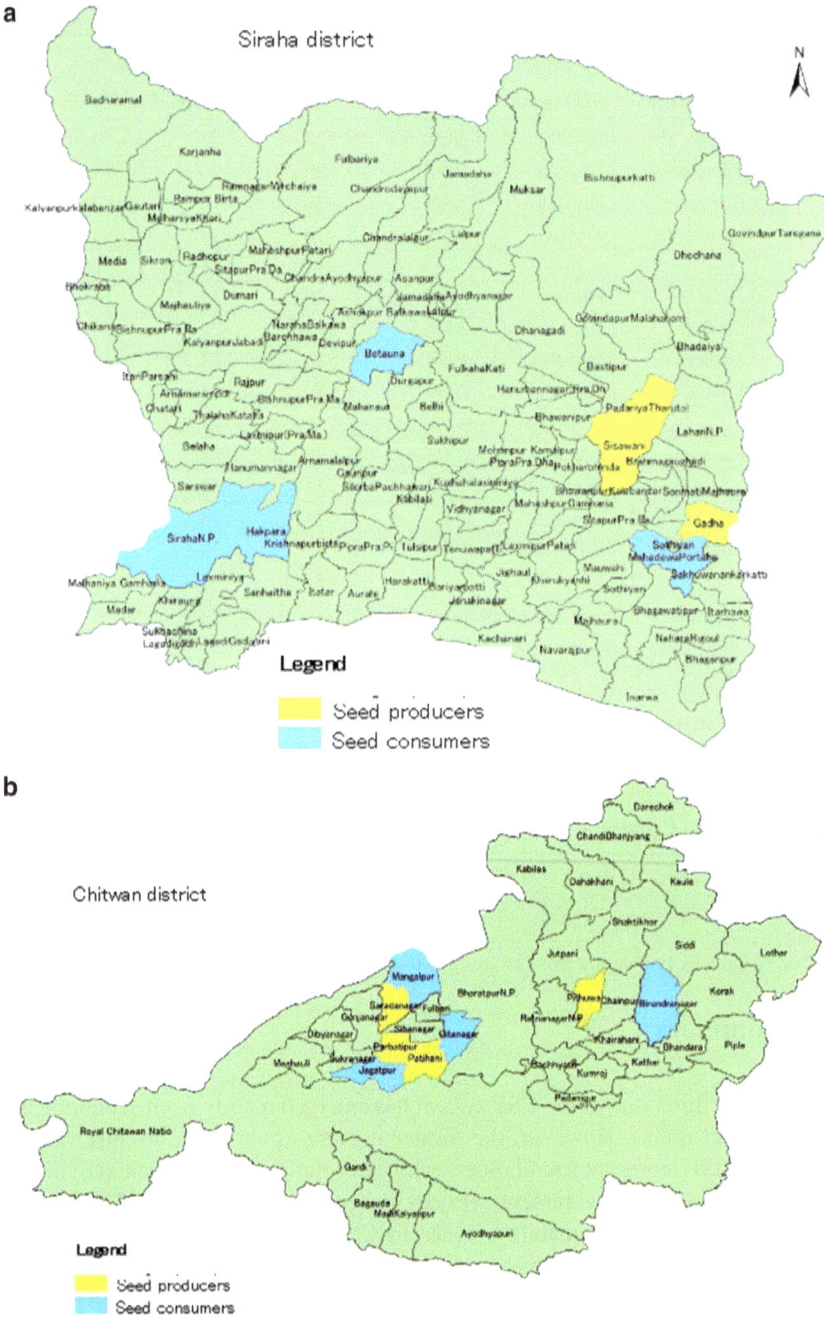

Fig. 3.4 (**a**) Surveyed village development communities (VDCs) in the selected districts (**b**) Surveyed VDCs in the selected districts (**c**) Surveyed VDCs in the selected districts (Source: Base map from ICIMOD.ORG)

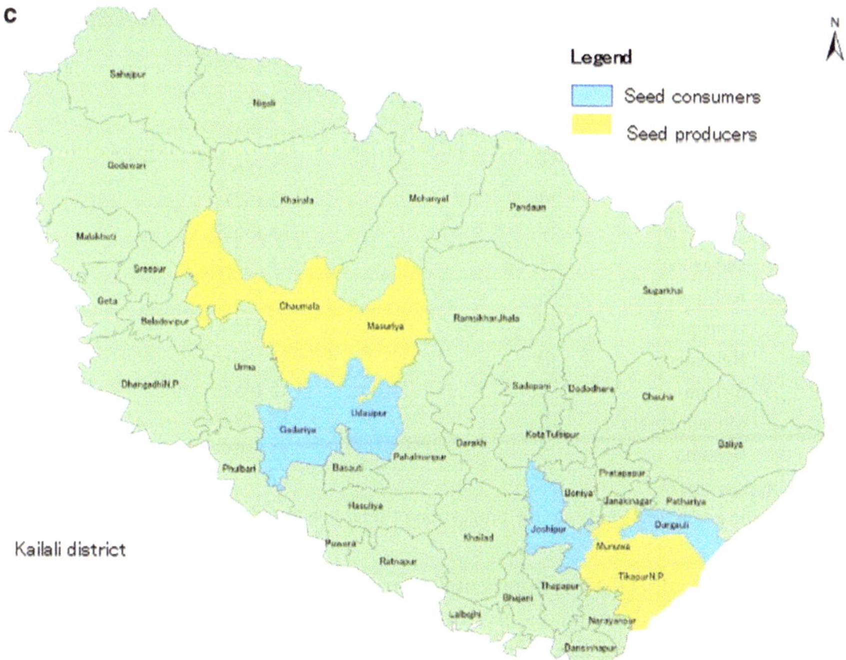

Fig. 3.4 (continued)

Fig. 3.5 Researchers discussing with seed consumers. (Source: Authors' photo bank)

Table 3.4 Profile of seed consumers selected for study in the area

District	VDC/ municipality	Total number of households	Households chosen for the study
Chitwan	Jagatpur	30	15 (50)
	Gitanagar	45	15 (33)
	Birendranagar	27	15 (56)
	Mangalpur	29	15 (52)
Kailali	Gadariya	27	15 (55.5)
	Durgauli	23	15 (65.2)
	Joshipur	28	15 (53.5)
	Udasipur	41	15 (36.5)
Siraha	Hakpada	30	15 (50)
	Sisbani	33	15 (45)
	Mahadevpratoha	19	15 (78.9)
	Betauna	25	15 (60)

Source: Survey 2011
Figures in the parentheses indicate the proportion of total households

production represents the environmental performance, whereas organizational governance captures both social and institutional concerns. Soil conservation practices together with risk management practices explain the continuity of current benefits in future, whereas economic benefit and organizational governance show the adaptive and innovation capacity of farmers. The framework discussed in this chapter is based on theoretical ideas and findings from previous studies. Thus, it is important to analyze the sustainability of CSP using empirical data.

References

Almekinders CJM, Louwaars NP, de Bruijn GH (1994) Local seed systems and their importance for an improved seed supply in developing countries. Euphytica 78:207–216

Bishaw Z, van Gastel AJG (2008) ICARDA's seed delivery approach in less favorable areas through village-based seed enterprises: conceptual and organizational issues. J New Seeds 9 (1):68–88

Brinkerhoff DW, Goldsmith AA (1990) Institutional sustainability in agriculture and rural development. Praeger, New York

CBS (2003) National sample census of agriculture Nepal, 2001/02; Highlights, Kathmandu, Nepal, National Planning Commission Secretariat, Central Bureau of Statistics

Chand SP, Karki M (2005) Community-based seed production in maize: present status and future opportunities. CIMMYT, Kathmandu

Cook ML (1995) The future of US agricultural cooperatives: a neo-institutional approach. Am J Agric Econ. Retrieved 10 Apr 2013, from http://web.missouri.edu/~cookml/publicat.htm-35k.

Crissman CC, Uquillas JE (1992) The seed sector in developing countries: a framework for performance analysis. Working paper 65. Overseas Development Institute, London

David S (2004) Farmer seed enterprises: a sustainable approach to seed delivery? J Agric Hum Val 21:387–397

FAO (1998) Developing seed security strategies and programs for food security in developing countries. In: Proceedings of the international workshop on seed security for food security: contribution for the development of seed security strategies in disaster prone regions Florence, 30 November–1 December 1997

FAO (2010) Promoting the growth and development of small seed enterprises for food security crops. Retrieved 10 Mar 2011, from www.fao.org/docrep/013/i1839e/i1839e00.pdf

Farrell MJ (1957) The measurement of productivity efficiency. J R Stat Soc Ser 120:253–290

FORWARD (2013) Annual report. Bharatpur, Chitwan

Gray TW, Kraenzle CA (1998) Member participation in agriculture cooperative: a regression and scale analysis. Retrieved 10 Apr 2011, from www.rurdev.usda.gov/rbs/pub/rr165.pdf

Hunington SP (1968) Political order in changing society. Yale University Press, New Haven

Idiong IC (2007) Estimation of farmer level technical efficiency in small scale swamp rice production in cross river state, Nigeria: a stochastic frontier approach. World Agric Sci 4:45–60

Joshi KD, Devkota KP, Harris D, Khanal NP, Paudyal B, Sapkota A, Witcombe JR (2012) Participatory approaches rapidly improve household food security in Nepal and identify policy changes required for institutionalization. Field Crop Res 131:40–48

Khanal NP, Maharjan KL (2010) Sustainability of community-based seed production enterprises in Nepal: institutional issues. Nepal Agric Res J 10:33–40. http://pustakalaya.org/eserv.php?pid=Pustakalaya:3772&dsID=NARC2010_NepalAgricultureResearchJournalVol10.pdf

Khanal NP, Maharjan KL (2013) Factors influencing households' behavior in selling rice seed in the market. Agric Food Econ 1–14 http://www.agrifoodecon.com/content/1/1/14

Lal KK, Thapa M, Guenat D (2009) Review of seed projects in Nepal. Swiss Agency for Development and Cooperation, Kathmandu

Maharjan KL (2003) Peasantry in Nepal: a study on subsistence farmers and their activities pertaining to food security. Research Center for Regional Geography, Hiroshima University, Hiroshima

MoAC (2013) Statistical information on Nepalese agriculture. Ministry of Agriculture and Cooperatives, Kathmandu

Morris ML, Rusike J, Smale M (1998) Maize seed industries: a conceptual framework. In: Morris ML (ed) Maize seed industries in developing countries. Lynne Rienner, Boulder

Nkonya E, Norman D (1997) Factors affecting adoption of improved maize in Northern Tanzania. Retrieved 30 Aug 2012, from http://onlinelibrary.wiley.com/doi/10.1111/j.1477-9552.1997.tb01126.x/pdf.

Onions CT (1964) The shorter Oxford English dictionary. Clarendon Press, Oxford

Paudel P, Matsuoka A (2008) Factors influencing adoption of improved maize varieties in Nepal: a case study from Chitwan district. Aust J Basic Appl Sci 2(4):823–834

Piya S, Kiminami A, Yagi H (2012) Comparing the technical efficiency of rice farmers in urban and rural areas: a case study from Nepal. Trends Agric Econ 5:48–60

Pokhrel S (2012) Role of DISSPRO and CBSPs in current seed supply situation in Nepal. J Agric Environ 13:53–59

Poudel D, Choudhary P, Chowin KR, Ghimire H (2003) Seed production and marketing through farmers' groups: case studies. Center for Arid Zone Studies, University of Wales, Cardiff

Pradhan KK (2009) Rural finance in Nepal: prospective from cooperative sector. Cooperative Development Board, Kathmandu, Nepal. Sahakari Sambad Ann Bull 46–56

Rana RB, Garforth C, Stapit B, Jarvis D (2007) Influence of socio-economic and cultural factors in rice variety diversity management on-farm in Nepal. Agric Hum Values 24:461–472

SQCC (2013) Seed vision. Seed Quality Control Center, Ministry of Agriculture and Cooperative, Kathmandu

Tripp R (2001) Provision of seed for agricultural development. Retrieved 22 Apr 2011, from http://mcknight.ccrp.cornell.edu/program_docs/litlists/Rach_Biblio_Seed%20Syst_18may06.pdf

Tripp R, Walker D, Delimini LL (1998) Seed management by small scale farmers in Zambia. Retrieved April 2, 2011, from www.informaworld.com/index/903754937.pdf

UNDP (2014) Nepal human development report. Retrieved 22 Jul 2014, from www.np.undp.org

WECD (1987) From one earth to one world: an overview. World Commission on Environment and Development. Retrieved 10 Apr 2011, from www.wsu.edu/~susdev/WCED87.html-5k

White LG (1984) Lessons in development project management. University of Maryland, College Park, Maryland

Witcombe JR, Virk DS (1997) New directions for public sector variety testing. In: Tripp R (ed) New seed and old laws regulatory reform and the diversification of national seed systems. Overseas Development Institute, London

Witcombe JR, Devkota KP, Joshi KD (2010) Linking community-based seed producers to markets for a sustainable seed supply system. J Exp Agric 46(4):425–437

Chapter 4
Farmers' Behavior in Buying Rice and Wheat Seed from Market

Abstract The market has a significant role in the success of any economic intervention, including seed production. So, it is important to analyze the characteristics of potential consumers and their behavior in buying seed from markets while assessing the performance of community seed production. This chapter analyzes the impact of socioeconomic characteristics in household behavior in buying rice and wheat seed. A set of demographic, economic, and institutional variables was tested for their impact on the probability of buying rice and wheat seed from markets by using a binary logistic model. Data obtained from 180 households spread across the three Tarai districts of Nepal were used in the analysis. Results show that 72 % of households buy rice seed from the market whereas 62 % buy wheat seed. Most households buy at least two rice varieties but only one variety of wheat. Household behavior in buying seed is mainly influenced by education, operational holding, irrigation facility, household membership in farmers' groups and cooperatives, and seed price.

Keywords Socioeconomic characteristics • Farmers' behavior • Rice and wheat seed • Market

4.1 Introduction

Seed is the most important input in agriculture, and its quality determines crop production potential and efficiency of other inputs used in crop fields (Almekinders et al. 1994). Considering this, research and development agencies motivate farmers for using quality seed as the potential of seed is likely to deteriorate over time because of physical admixture and genetic erosion. On the other hand, motivating farmers in buying seeds is important for seed producers to enhance their economic benefits as the buyers' behavior determine seed consumption. Farmers buy seed from the market for three principal reasons: to receive a new variety, to get quality seed of existing varieties, and to receive genetic material when they lose seed stock because of such disasters such as fire, flood, and drought. However, the first reason, a new variety, is more common (Gauchan et al. 2005; Rana et al. 2007). In spite of the benefits of growing quality seeds, farmers do not frequently buy seed from markets, especially in open-pollinated crops including rice and wheat. Rather, part

N.P. Khanal, K.L. Maharjan, *Community Seed Production Sustainability in Rice–Wheat Farming*, DOI 10.1007/978-4-431-55474-5_4

of the crop harvest will be saved for next year's planting, or households may receive seed from different informal networks such as friends, relatives, and neighbors.

Farmers' behavior in buying seeds can be discussed from the perspective of economic benefits. However, understanding those economic benefits is not simple in subsistence farming because crop yield is not the only determinant explaining benefits from the adoption of new seed, because straw yield and its quality, suitability of seeds in the existing cropping systems, and tolerance to biotic and abiotic stresses might be important in choosing new seeds (Joshi et al. 2001; Khanal and Maharjan 2014). In irrigated upland domains of the rice-wheat system, farmers could choose early-maturing rice varieties that do not produce as much as crop yield as the medium- or late-maturing types (Yadav et al. 2005; Kafle et al. 2012; Khanal and Maharjan 2014) but contribute to maximizing their benefits from the overall cropping system perspective. Growing early-maturing rice varieties allows farmers to grow wheat in a timeframe that increases its yield by minimizing terminal drought during the milking stage (Regmi et al. 2009). This arrangement also encourages farmers to grow spring season crops, such as vegetables, after the wheat harvest (Khanal and Maharjan 2014). Moreover, the valuation methods and level of benefits farmers realize from new seed also vary with demographic factors, resource endowments, and institutional variables (Nkonya and Norman 1997; Sheikh et al. 2003; Joshi and Bauer 2006; Rana et al. 2007; Paudel and Matsuoka 2008; Khanal and Maharjan 2014). The aforementioned factors make it complex to measure farmers' behavior in buying seed directly. This chapter analyzes households' behavior in buying rice and wheat seed from the market indirectly from a microeconomic perspective.

4.2 Methodology

4.2.1 Site Selection and Sampling Technique

The study was carried out in Siraha, Chitwan, and Kailali districts representing the Tarai region of Nepal. This region contributes more than 60 % of the total rice-wheat system of Nepal (MoAC 2011). In each district, four village development committees (VDCs), the lowest administrative unit, were selected for household survey. These VDCs include Jagatpur, Mangalpur, Gitanagar, and Birendranagar municipality (Chitwan district); Hakpada, Sisbani, Mahadevparoha, and Betauna (Siraha); and Gardaiya, Joshipur, Durgauli, and Udasipur (Kailali). One ward (of nine wards) from each VDC, and 15 households from each ward, were randomly selected, making 180 households for field survey. The survey was carried out using a semi-structured questionnaire tested in households not involved in the survey. One group discussion in each ward was organized after the completion of the household-level survey to collect additional necessary information.

4.2.2 Empirical Model

A binary logistic model (BLM) was used to analyze the impact of socioeconomic variables on farmers' behavior in buying rice and wheat seed. In this case, the dependent variable is binary, that is, it is 1 if farmers buy seed from the market and 0 otherwise. Although linear probability models such as ordinary least squares (OLS) can be used to analyze binary choice data, certain assumptions of classical regressions such as non-normality and heteroscedastic error, and questionable R^2 as a measure of goodness of fit, are violated. Logit and probit models have been developed to address these issues; however, the logit model is preferred if the choice variables are mutually exclusive with each other (Long and Freese 2006). Previous researchers (Joshi and Bauer 2006; Paudel and Matsuoka 2008) also adopted a BLM to analyze farmers' behavior in buying seed. Theoretically, the BLM is given in Eq. 4.1 (Agresti 1996):

$$\mathrm{Ln}(P_x/(1-P_x)) = \beta_0 + \beta_1 X_i + \beta_2 X_i + \ldots\ldots.\beta_k X_i + \varepsilon_i \tag{4.1}$$

where Ln is log, i is the ith observation in the sample, P_x is the probability of farmers buying seed from the market in consideration of the given explanatory variables ($X_{i,}$) and ($1-P_x$) is the probability of non-adoption. β_0 is the coefficient of intercept, and β_1, β_2........ β_k are parameters to be estimated, k indicates the types of explanatory variables, and ε_i is error term. Because the BLM is estimated through a maximum likelihood method, the coefficients do not show the average impact of independent variables on the probability of buying seed from the market. So, the marginal effect of socioeconomic variables on dependent variables was estimated after estimating the BLM (Sheikh et al. 2003). The marginal effect values are used to discuss the influence of explanatory variables on households' probability of buying seed.

Specification of the Model and Variables

With reference to the theoretical model given in Eq. 4.1, the operational model used in the study is specified in Eq. 4.2.

$$\begin{aligned} Y = \beta_0 &+ \beta_1 \text{ ln age of HHH} + \beta_2 \text{ ln education of HHH} \\ &+ \beta_3 \text{ ln family labor} + \beta_4 \text{ ln off-farm income} \\ &+ \beta_5 \text{ ln operational land} + \beta_6 \text{ ln livestock} \\ &+ \beta_7 \text{ ln chemical fertilizer} + \beta_8 \text{ irrigation} + \beta_9 \text{ CBO} \\ &+ \beta_{10} \text{ ln seed price} + \beta_{11} \text{ Chitwan} \end{aligned} \tag{4.2}$$

Here, Y represents the binary dependent variable (0, 1). Explanatory variables were selected considering production theory, previous studies, and the field situation. These variables are classified as demographic [age and education of household

Table 4.1 Socioeconomic variables included in binary logistic regression

Variables	Definition	Expected sign
Age of head of household (HHH)	Age of household head (years)	–
Education of HHH	Formal education of HHH (years of schooling)	+
Family labor	Labor force unit (LFU)[a] at household	+
Off-farm income	Household annual cash income from off-farm sources (NRs)	+
Operational land	Operational land for crop production (ha)	+
Livestock	Livestock standard unit (LSU)[b] at household	+
Chemical fertilizer	Total cost of chemical fertilizer (NRs ha^{-1})	+
Irrigation	1 if farmers have access to public irrigation facility, and 0 otherwise	+
CBOs	1 if any member of household has membership in farmer group/cooperative, and 0 otherwise	+
Seed price	Price of seed (NRs kg^{-1})	–
Chitwan	1 if farmers from Chitwan district, and 0 otherwise	+

CBOs community-based organizations, *NRs* Nepalese rupees, *1 US$* NRs 82.96
[a]LFU is the measurement of labor force, where people from 15 to 59 years old regardless of their sex were categorized as 1 person = 1 LFU; in case of children (10–14 years old) and elderly people (>59 years old), 1 person = 0.5 LFU
[b]LSU is the aggregates of different types of livestock kept at a household in standard units calculated using the following equivalents: 1 adult buffalo = 1 LSU, 1 immature buffalo = 0.5 LSU, 1 cow = 0.8 LSU, 1 calf = 0.4 LSU, 1 pig = 0.3 LSU, 1 sheep or goat = 0.2 LSU, 1 poultry or pigeon = 0.1 LSU (Khanal and Maharjan 2014)

head (HHH) and family labor), economic (off-farm income, operational land, livestock, chemical fertilizer, irrigation, seed price); and institutional [household's membership in community-based organizations CBOs)]. The description of the explanatory variables and their hypothesized influence on a household's seed-buying behavior is summarized in Table 4.1.

Commonly, the HHH is the major decision maker of the household, and therefore its characteristics might influence the household decision in buying rice and wheat seeds. It was hypothesized that a younger HHH might be better off in buying seed from market as compared to an older HHH because of having a better linkage in the market (Paudel and Matsuoka 2008). It was hypothesized that the education of HHHs might have a positive impact on household seed-buying behavior as it contributes to innovativeness and creativity. Higher-educated people could establish a better linkage with research and extension agencies, and these agencies are the major sources of accessing training and inputs for crop production (Sheikh et al. 2003; Joshi and Bauer 2006). Similarly, family labor is an important input in subsistence farming, and it was hypothesized to have a positive influence on household seed-buying behavior. Households with more family labor might have motivation to increase crop yield by integrating different inputs including labor. Availability of family labor means that households are more likely to implement field activities on time (planting, weeding, fertilizer application), and it motivates farmers to buy seed (Paudel and Matsuoka 2008).

Among the economic variables, a household's off-farm cash income was assumed to have a positive impact on seed-buying behavior: access to credit is still challenging in the study area as most of the banks and micro-finance institutions are located in the cities, and it is difficult for small farmers in these areas to access credit (Pradhan 2009). It was hypothesized that those having access to off-farm income could more easily get cash for buying necessary inputs for rice and wheat production and implementing crop husbandry activities in time. Similarly, operational landholding was considered to have a positive influence considering their motivation to increase the yield by combining other agricultural inputs (Tiwari et al. 2008). Livestock is the integral component of the Nepalese farming system, and farmers use all the manure, whatever is produced at the household, on their fields. Hence, livestock standard unit was calculated as a proxy variable to represent the amount of animal manure applied in rice and wheat and is assumed to have a positive influence on buying seed for these crops. In addition to animal manure, farmers also use chemical fertilizers in rice and wheat production, and this was hypothesized to have a positive impact on a household's seed-buying behavior considering its positive linkage with crop yield (Paudel and Matsuoka 2008; Regmi et al. 2009).

Similarly, it was hypothesized that access to irrigation could have a positive influence on a household's seed buying as it has also a positive link with crop yield. However, seed price was hypothesized to have a negative influence on seed buying considering an inverse relationship between price and demand.

The institutional variable included in this study is the household's membership in an agricultural group or cooperative (referred to as community-based organizations, or CBOs, in this chapter), and it was considered to have a positive impact on seed-buying behavior. Those farmers having membership in CBOs are more likely to access extension facility including training from these agencies in those areas where the agricultural extension system is focused on the group approach (Tiwari et al. 2008). Moreover, Chitwan district has better road networks, infrastructure, and extension facility as compared to the other two districts. These factors would be additional sources of variation in the model and could distort the result. To address this issue, Chitwan (location) was used as a dummy variable while running the model.

As per the regression rule, diagnostic tests were carried out to check the heteroscedasticity and multicollinearity problem in the data. For this, selected explanatory variables were regressed against the dependent binary variable using the ordinary least squares (OLS) technique. The variation inflation factor (VIF) test was carried out to check multicollinearity among the variables. The VIF value for the dependent variables remained below 10 , suggesting no problem of multicollinearity. The Breusch-Pagan/Cook-Weisberg test was carried out to test for the heteroscedasticity, and the null hypothesis of heteroscedasticity was strongly rejected (p value less than 0.004).

4.3 Results and Discussion

4.3.1 Description of Study Variables

The study shows that for rice 72 % of the households buy seed from the market, and the seed comprises 17 varieties. These varieties are both modern varieties (58.8 %) and farmers' varieties (41.2 %). Of the total modern varieties (10), only 50 % of them (Savitri, Hardinath 1, Ramdhan, Mithila, and Radha 4) were released by the Nepalese government, whereas the other 4 modern varieties (Kanchhi Masuli, Sarju 52, Sona Masuli, and Sawa Masuli) are those released by the Indian government in the 1970s and introduced in Nepal's Tarai districts through informal channels (farmer-to-farmer contact) because of the open-border system between Nepal and India. However, 63 % of the households buy wheat seed from the market. It was found that farmers bought a total of seven modern varieties, all of which have been released by the Nepalese government (Nepal Agriculture Research Council; NARC 2011). Among these varieties, Gautam (65 %) remains the most popular wheat variety, followed by NL 297 (50 %) and BL 1473 (25 %). In contrast to rice, wheat seed is exported to India from Nepal through an informal way. Households buy rice and wheat varieties from the market from various sources, but neighboring farmers, agrovets (local traders), farmers' groups, and agricultural stations are used, in order of importance. There are also regional differences in the types of varieties bought by farmers. For example, Hardinath 1, Ramdhan, and Radha 4 have wider adaptation across the surveyed districts whereas other rice varieties have niche specificity. For instance, Savitri in Chitwan, Sarju 52 in Kailali, and Kanchhi Masuli in Siraha are more popular than other varieties grown in these locations. There is also variation in number of varieties bought by farmers. For rice, 83 % of households buy seeds of at least two varieties, and 17 % grow one rice variety. However, wheat is almost the opposite: 78 % of households grow only one variety (Fig. 4.1).

Table 4.2 shows the summary statistics of socioeconomic variables of the households included in the binary logistics model with reference to their mean and standard deviation. The average age of the HHH is 41.9 years, but it varies from

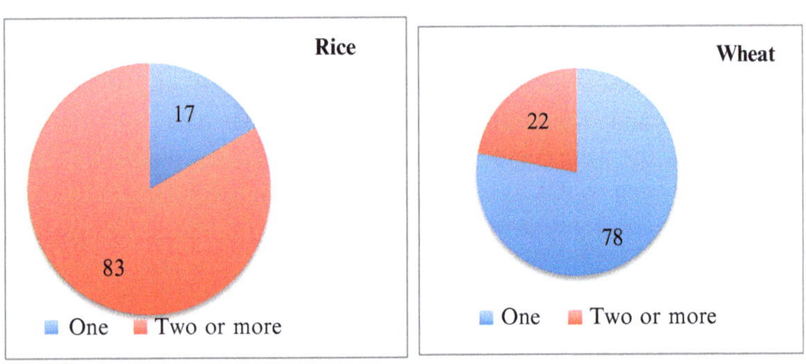

Fig. 4.1 Farmers' behavior in buying rice and wheat seed from the market (From Survey 2011)

Table 4.2 Summary of socioeconomic variables included in binary logistic model

Variables	Overall mean	Chitwan	Siraha	Kailali
Age of HHH	41.9 ± 13.64[a]	49.28 ± 13.48	42.03 ± 12.50	34.34 ± 10.63
Education of HHH	5.20 ± 1.58	6.6 ± 6.61	4.81 ± 5.42	4.08 ± 5.12
Family labor	3.2 ± 2.76	3.6 ± 2.10	3.10 ± 2.32	2.80 ± 0.78
Off-farm income	49,531 ± 67,890	68,640 ± 42,580	48,875 ± 32,256	37,815 ± 20,452
Operational land (rice)	0.78 ± 0.66	0.58 ± 0.45	0.86 ± 0.66	0.91 ± 0.78
Operational land (wheat)	0.58 ± 0.35	0.34 ± 0.41	0.67 ± 0.48	0.78 ± 0.57
Livestock	3.46 ± 1.85	5.06 ± 3.56	1.49 ± 0.48	2.85 ± 1.45
Chemical fertilizer (rice)	5,244 ± 1,245	3,594 ± 1,721	5,530 ± 1,493	6,654 ± 3,298
Chemical fertilizer (wheat)	5,405 ± 2,646	3,815 ± 2,122	5,680 ± 2, 026	6,720 ± 3,765
Irrigation	0.34 ± 0.47	0.39 ± 0.49	0.36 ± 0.48	0.26 ± 0.44
CBO	0.56 ± 0.23	0.68 ± 0.24	0.48 ± 0.21	0.51 ± 0.34
Seed price (ice)	20.5 ± 16.7	21.3 ± 10.8	20.8 ± 14.21	19.4 ± 8.79
Seed price (wheat)	21.75 ± 14.3	22.4 ± 11.2	21.6 ± 9.4	20.3 ± 12.6
n	180	60	60	60

Source: Survey 2011
[a]Standard deviation

17 to 74 years. The majority (96 %) of HHHs have attended formal schooling, and their average years of schooling are just 5.2 (primary education in the Nepalese standard). The average labor force of a household is 3.2 persons, but this number varies from 2 to 15. Similarly, the average operational holding per household is 0.78 ha (range, 0.06–4.67 ha), which is similar to the national average operational holding (0.8 ha). Household annual cash income is NRs 59,922, but only 63.3 % of households receive cash income from agriculture. Two-thirds of the households (66.67 %) receive cash income from non-farm sources such as a salaried job, remittance, wage labor, or small business. The average non-farm income of households is NRs 49,531, varying from NRs 4,780 to NRs 122,600. Livestock is an integral part of farming system in the study area. All the households were found to have raised livestock, and the average livestock unit (LSU) in the study area is 3.46, varying from 0.5 to 201. Cow, buffalo, goat, poultry, and pig are the major livestock species being raised by farmers.

In addition to animal manure, 90 % of farmers apply chemical fertilizers in rice and wheat. The sources of chemical fertilizers are urea (nitrogenous fertilizer, 60 % N), diammonium phosphate (18 % N and 48 % P), and muriate of potash (60 % K). It was found that the ratio of chemical fertilizers applied by farmers in rice fields is N:P:K (44.9:25:20.9 kg ha^{-1}), and this dose is smaller than the recommendation made for irrigated rice in the Tarai region of Nepal of 100:30:30 kg ha^{-1}. Fertilizer application is also similar in wheat where farmers apply N:P:K (50.9:22:17.9 kg ha^{-1}), and it is also lower than the recommended fertilizer dose for irrigated wheat (MoAC 2011). Farmers apply both potassium and phosphorus as basal applications during final land preparation, whereas nitrogen (urea) is applied 30 to 50 days after planting.

Chemical fertilizer cost (NRs) was used to represent the amounts of chemical fertilizer applied in rice and wheat crops. On average, farmers apply chemical fertilizer at a cost of NRs 5,244 ha^{-1} and NRs 5,405 ha^{-1} in rice and wheat, respectively. Sampled households use irrigation from both public irrigation sources (such as canal from river or stream) or private irrigation sources (tube well). However, only 34 % of the households have access to public irrigation source. Similarly, 56 % of the households have membership in CBOs.

4.3.2 Result of Binary Logistic Model

The significant log-likelihood statistic shows that the variables chosen for the study fit the model well (Tables 4.3 and 4.4); that is, the coefficients of explanatory variables are significantly different from zero. Moreover, the percentage correctly predicted from the model is also high (74.5 % for rice and 68.4 % for wheat). These results show the goodness of fit of the model in the given data set. The study shows that the coefficients of most of the independent variables are as hypothesized; however, there is some variation.

Table 4.3 Impact of socioeconomic variables on farmers' behavior in buying rice seed

Variables	β	*P* value	Marginal effect[a]
Age of HHH	0.031	0.184	0.0041
Education of HHH	0.112	0.121	0.013
Family labor	−0.043	0.165	0.117
Off-farm income	−0.0031	0.243	0.0001
Operational land	0.811	0.124	0.140
Livestock	0.0027	0.943	0.0003
Chemical fertilizer	0.0002	0.705	0.0002
Irrigation	0.812	0.03***	0.301
CBO	0.641	0.079*	0.127
Seed price	−0.240	0.0127**	0.014
Chitwan	0.221	0.014**	0.125
Constant	−2.184	0.012	

Log likelihood, 85.37**, $n = 107$; percentage correctly predicted, 74.5, pseudo R^2, 0.22
*, **, *** indicate significance at 10 %, 5 %, and 1 % levels, respectively
[a]Marginal effect = $p\ (1-p)\beta$, where p is the probability of the event occurring, and β is the parameter estimated from the model

Table 4.4 Impact of socioeconomic variables on farmers' behavior in buying wheat seed

Variables	β	*p* value	Marginal effect[a]
Age of HHH	−0.002	0.952	−0.0006
Education of HHH	0.177	0.097*	0.043
Family labor	−0.018	0.528	−0.0045
Off-farm income	−0.001	0.233	−0.001
Operational land	1.58	0.032**	0.3902
Livestock	0.395	0.324	0.0973
Chemical fertilizer	0.003	0.415	0.0009
Irrigation	1.70	0.075*	0.401
CBO	1.24	0.021**	0.352
Seed price	0.35	0.024**	0.214
Chitwan	−1.72	0.277	−0.376
Constant	−3.23	0.235	

Log likelihood, −22.54**, $n = 115$; percentage correctly predicted, 68.4; pseudo R^2, 0.45
*, **, *** indicate significance at 10 %, 5 %, and 1 % levels, respectively
[a]Marginal effect = $p\ (1-p)\beta$, where p is the probability of the event occurring, and β is the parameter estimated from the model

Among the explanatory variables, irrigation, household membership in CBOs, and seed price have shown significant influence on household decisions to buy seed from the market. Households that access an irrigation facility realize a rice yield 12.4 % higher than their counterparts (2.42 t ha^{-1}). Similarly, in wheat crops, those accessing irrigation facilities experienced higher grain yield (4.7 %) than those who do not have irrigation access (1,879.8 kg ha^{-1}).

In this study, household membership in CBOs represents access to extension facilities (e.g., agricultural training) of government and nongovernment organizations. The significant coefficient and high marginal effect signify the influence of CBO membership on household behavior in buying both rice and wheat seed from the market. As shown in Table 4.3, the marginal effect of households having membership in CBOs is 0.127, which indicates that households having membership in CBOs have a 12.7 % higher probability of buying rice seed from market than those not having membership. The influence of CBO membership is also positive, and even higher in wheat. Households having CBO membership have 35.2 % higher probability of buying wheat seed from the market. This finding is consistent with other studies (Paudel and Matsuoka 2008; Tiwari et al. 2008). Associated reasons might be better access to agricultural training and other technical facilitations households receive from extension agencies (DFID 2010). In the group discussions, farmers opined that as a member of the CBO households participate in the monthly meeting and demonstration plots. They also discuss the problems, lessons, and potentials of new crop varieties and technologies in monthly meetings. All these factors might have influenced adoption decisions.

This study also shows that the price of seed is significant in a household's decision to buy seed from the market; that is, households encountering a higher seed price are less likely to buy seed from the market. Thus, it is necessary to supply rice and wheat varieties at a reasonable price to enhance their motivation in buying seed. The essence of reasonable price here is seed price comparable with grain price of the respective commodity.

Farmers opined that normally seed price remains 15–20 % higher than grain price both in rice and in wheat. This small price difference between seed and grain might be because major suppliers in rice (93 %) and wheat (82 %) seed are neighboring farmers. Previous studies have shown that farmers normally compare seed price with grain price in open-pollinated crops because they could use home-saved seed if the price difference of these two categories of seed is high (Almekinders et al. 1994).

In addition, education of the HHH and operational land have also shown significant positive impact on household behavior in buying seed. Household response toward these variables is also in the same direction in rice, although nonsignificant. The positive impact of education is linked with household creativity and extension contact. Similarly, the motivation of larger operational holdings in buying seed is also clear as per the hypothesis, and it might result from the higher commercial orientation of larger operated households in maximizing their benefits from crop production.

4.4 Conclusion

This chapter analyzed farmers' behavior to buy rice and wheat seed from the market. It was found that 72.1 % of farmers buy rice seed from the market but this proportion is only 63 % in wheat. The majority of farmers buy at least two rice varieties from the market but in wheat they buy only one variety at a time. Seed-buying data alone could not provide the total household variety diversity portfolio of these crops because farmers grow many other local varieties from households' saved seed. A binary logistic model was used to understand the impact of household-level characteristics on farmers' seed-buying behavior. The result clearly shows that irrigation facility, household membership in CBOs, and seed price have significant roles in motivating farmers to buy seed from the markets. Although an irrigation facility has a significant positive impact on household seed-buying behavior, it might be a challenging task for governments and extension agencies to support this facility considering their resource constraints. Development of drought-tolerant varieties could be an alternative strategy in such cases. This study also suggests motivating farmers to actively engage in groups and cooperatives because, despite the significant positive influence of household membership in a CBO, about half the households are not engaged in a CBO. Moreover, this study shows that households experiencing higher price of seed are less likely to adopt improved varieties. Thus, emphasis should be placed on rice seed production at local levels so that cost of production and transportation could be reduced, thereby enhancing accessibility of cheaper rice and wheat seeds to the farmers.

References

Agresti A (1996) An introduction to categorical data analysis. Wiley, New York

Almekinders C, Louwaars NP, Bruijin GH (1994) Local seed systems and their importance for an improved seed supply in developing countries. Euphytica 78:207–216

DFID (2010) Uptake of new rice varieties in India. The impact assessment report. Department for International Development, Kathmandu

Gauchan DM, Smale M, Chaudhary P (2005) Market-based incentive for conserving diversity on-farm: the case of rice landraces in central Tarai, Nepal. Genet Res Crop Evol 52:293–303

Joshi G, Bauer S (2006) Farmers' choice of modern rice varieties in the rainfed ecosystem of Nepal. J Agric Rural Dev Trop Subtrop 107(2):129–138

Joshi KD, Sthapit B, Witcombe JR (2001) How narrowly adopted are the products of decentralized breeding? The spread of rice varieties from a participatory plant breeding program in Nepal. Euphytica 122:589–597

Kafle B, Paudel MN, Ghimire RC (2012) Assessing current status of rice varieties in river basin area of Nepal: concern for diffusion. Int J Agric Res Rev 2(2):59–61. Retrieved 10 July 2012, from Electronic Center for International Scientific Information: http://www.ecisi.com

Khanal NP, Maharjan K (2014) Adoption of improved rice varieties in the Tarai region of Nepal. In: Maharjan KL (ed) Communities and livelihood strategies in developing countries. Springer, Hiroshima, pp 49–64. doi:10.1007/798-4-431-54774-7-4

Long ST, Freese J (2006) Regression model for categorical dependent variables using stata. Stata Press, College Station
MoAC (2011) Statistical information on Nepalese agriculture. Agri-business Promotion and Statistical Division, Kathmandu
NARC (2011) Released and registered crop varieties in Nepal (1960–2010). Retrieved July 15, 2012, from http://narc.gov.np/publicaton/pdf/varieties_released/Recorded%20released%20varieties.pdf
Nkonya E, Norman D (1997) Factors affecting adoption of improved maize in Northern Tanzania. Retrieved 30 Aug 2012, from Wiley online library: http://onlinelibrary.wiley.com/doi/10.1111/j.1477-9552.1997.tb01126.x/pdf
Paudel P, Matsuoka A (2008) Factors influencing adoption of improved maize varieties in Nepal: a case study from Chitwan district. Aust J Basic Appl Sci 2(4):823–834
Pradhan KK (2009) Rural finance in Nepal: prospective from cooperative sector. Cooperative Development Board, Kathmandu. Sahakari Sambad Ann Bull
Rana RB, Garforth C, Stapit B, Jarvis D (2007) Influence of socio-economic and cultural factors in rice variety diversity management on-farm in Nepal. Agric Hum Values 24:461–472
Regmi A, Tripathi G, Giri S, Bhatta M, Sherchan DP, Karki KB, Kumar V, Ladha JK (2009) Improving food security through integrated crop and resource management in the rice-wheat system in Nepal. Retrieved 25 Sept 2011, from: International Rice Research Institute: http://irri.org/knowledge/publications/irri-books/integrated-crop-and-resource-management-in-the-rice-wheat-system-of-south-asia
Sheikh AD, Rehman T, Yates CM (2003) Logit models for identifying the factors that influence the uptake of new technologies by farmers in the rice-wheat and the cotton-wheat farming systems of Pakistan's Punjab. Agric Syst 75:79–95
Tiwari KR, Sitaula BK, Nyborg ILP, Paudel GS (2008) Determinants of farmers' adoption of improved soil conservation technologies in mid mountain watershed of central Nepal. Environ Manag 42:210–222
Yadav RB, Chaudhary B, Adhikari NP, Akhtar T (2005) Hardinath 1, an early maturing rice genotype released in Nepal. Retrieved 11 Jul 2012, from http://dspace.irri.org:8080/dspace/bitstream/123456789/1144/1/Yadaw,%20R.%20B.%20Hardinath%201,%20an%20early.pdf

Chapter 5
Technical Efficiency in Rice and Wheat Seed Production

Abstract Measuring technical efficiency is a popular approach to understand farmers' performance in seed production. This chapter measures farmers' technical efficiency (TE) in utilizing seed, chemical fertilizer, labor, livestock, and operational land for producing rice and wheat seed using survey data collected from 180 households across the three Tarai districts of Nepal. The stochastic frontier production model was used for data analysis as it gives us better estimates of efficiency in utilizing the aforementioned resources by removing measurement errors. Results show that farmers are 81 % efficient in utilizing resources; this ranges from 36.7 to 95.2 %. They have also similar technical efficiency level (78.3 %) in wheat seed production, and there is wide variation in efficiency (38.6–94.6 %) across the selected households. The differences in efficiency level are explained by the education of household head, land rent, and farmers' experience in seed production. This study recommends provision of irrigation facility, soil measurement methods, and vocational training to enhance farmers' efficiency in rice and wheat seed production.

Keywords Rice-wheat seed production • Technical efficiency • Stochastic frontier model

5.1 Introduction

The term efficiency shows the ratio between output and input. This term has been popular in the literature since Farrell (1957) elaborated ideas about technical efficiency (TE) and allocative efficiency (AE). In the frontier concept, a firm is said to be technically efficient if its output falls from possible maximum output level in the given set of inputs, and similarly a firm is said to be allocatively efficient if it applies the inputs in appropriate proportion (by equating the ratio of marginal product of input with input price ratio) in the observed input price and output level (Battese and Coelli 1995). Farmers' efficiency in resource mobilization can be measured from production (TE), cost (AE), or profit (profit efficiency, PE) functions. Measuring efficiency is a popular approach to understand farmers' performance and, if inefficiency exists, it forms a basis for designing support strategies

N.P. Khanal, K.L. Maharjan, *Community Seed Production Sustainability in Rice–Wheat Farming*, DOI 10.1007/978-4-431-55474-5_5

(World Bank 2011). The gain in efficiency would contribute to transforming the seed production system from subsistence to commercial (Morris et al. 1998).

This concept has resulted in a number of past studies in efficiency, and these studies have rejected Schultz's hypothesis (Schultz 1964) that poor farmers in developing countries are efficient in utilizing their resources (Kalirajan 1999; Rahman 2003; Hassan and Ahmad 2005; Ghaderzadeh and Rahimi 2008; Kamruzzaman and Islam 2008; Dung et al. 2011; Sohail et al. 2012; Piya et al. 2012). These studies have identified the existence of a wide range of efficiency among rural households, ranging from 12 % to 98 %. In a practical sense, it is very difficult to compare the efficiency level of farmers from one study to another because of variations in choosing input variables. This difficulty necessitates the measurement of efficiency at a local level using most commonly used input variables so that location-specific appropriate policy recommendation could be made. Thus, this chapter measures the technical efficiency of farmers in producing rice and wheat seed with the inputs used by farmers: source seed, chemical fertilizer, labor, livestock, and operational land, and identifies the socioeconomic variables influencing efficiency of farmers.

5.2 Methodology

5.2.1 Study Area and Sampling Technique

This study was carried out in three Tarai districts (Siraha, Chitwan, and Kailali) of Nepal in November 2011. The Tarai is the major rice and wheat production domain of Nepal. Four seed producer organizations (SPOs) with at least 2 years of experience in rice and wheat seed production from these districts were purposively selected. Fifteen households from each of the selected SPOs were chosen for the field survey, making a total of 180 sample households.

5.2.2 Empirical Technique

The stochastic frontier production model (SFPM) was used to analyze technical efficiency (TE). Two types of parametric frontier production functions are used in the literature in measuring the TE of farmers, deterministic and stochastic; however, the latter is considered more efficient because it considers two error terms, one of which separates the random noise effect from the total residual (also called composed error), and therefore the model gives a consistent estimate for efficiency/inefficiency (Battese and Coelli 1995). The theoretical idea of the SFPM is that no one can produce output beyond the theoretically possible limit. The measurement of efficiency/inefficiency is thus possible to show how far away from the theoretical limit are the agents. Aigner et al. (1977) and Meeusen and van den Broeck (1977) originally proposed this model, and its functional form is expressed as given in Eq. 5.1:

$$Y_i = f(x_i;\beta)\exp(v_i - u_i) \tag{5.1}$$

Here, Y_i is the quantity of production of the ith farm with i ranging from 1, 2. . .n, x_i is the explanatory variable input, β is the vector of parameters to be estimated, v_i represents the two-sided error term accounting for random variation in output caused by factors outside the control of farmers such as measurement errors, diseases, pest infestations in the fields, and natural calamities. Another term, u_i, represents the error term associated with farm-level technical inefficiency, and this inefficiency might occur from variation in socioeconomic variables such as education, extension, or infrastructure. Here, v_i is assumed to distribute independently of each u_i, and both these errors are supposed to be not correlated with explanatory variables (x_i). The noise component, v_i, is assumed to have zero mean and constant variance (σ_{xx}^2) and be distributed normally, whereas the inefficiency component u_i is assumed to have zero mean with variance (σ_u^2) and be distributed half normally. As proposed by Aigner et al. (1977), the log-likelihood function for the half-normal model is given in Eq. 5.2. This likelihood function estimates whether variation among observations is the result of inefficiency or occurred by chance. From the likelihood function (5.2), we get σ^2 and λ^2, where σ^2 is the total variation and is estimated by $\sigma^2 = \sigma_u^2 + \sigma_v^2$ and $\lambda^2 = \sigma_u{}^2|\sigma^2$. If $\lambda = 0$, there is no inefficiency effect, and the variation in the data is just the result of random noise, and the higher value of λ reflects a greater inefficiency effect, as explained by the model

$$\ln L(Y_i|\beta,\sigma\lambda) = -\frac{1}{2}\ln(\pi\sigma^2) + \sum_{(i=1)}^{n} \ln \phi\left\{\frac{-\varepsilon\lambda}{\sigma}\right\} - \frac{1}{2\sigma^2}\sum_{i=1}^{n} \varepsilon_i^2 \tag{5.2}$$

where $\varepsilon_i = v_i = u_i = \ln Y_i - x_i\beta$ is a composite error term and $\phi(x_i)$ is a cumulative distribution function of the standard normal variable evaluated at x_i. The TE of farmer$_i$ in the context of stochastic production function can be expressed as

$$\mathrm{TE}_i = Y_i/Y_i^* = f(x_i;\beta)\exp(v_i - u_i)/f(x_i;\beta)\exp(v_i) = \exp(-u_i) \tag{5.3}$$

where Y_i^* is the maximum possible output, and Y_i, x_i, β, v_i, and u_i are as already explained. Here, TE_i represents TE and is the ratio of farm output (crop yield in this research) relative to the maximum output that can be produced in the same level of input vectors. The value of TE_i ranges from 0 to 1. If $\mathrm{TE}_i = 1$, Y_i achieves the maximum value of $f(x_i;\beta)$, exp (v_i), and $\mathrm{TE}_i < 1$ represents the shortfall of production from the maximum possible production level in the environment characterized by stochastic elements that vary across farmers.

As mentioned already, this study has adopted the two-stage procedure in estimating the SFPM. In the first stage, TE was computed from the model using the variables of seed, labor, chemical fertilizer, livestock, and operational land. In the second step, TE (as dependent variable) was regressed against socioeconomic variables of farmers or farm using ordinary least squares, as this technique can be applied if the efficiency values are greater than 0 and less than 1. Previous studies have also adopted the two-stage procedures to analyze the efficiency of farmers in developing countries (Kalirajan 1999; Sharma et al. 1999; Piya et al. 2012).

After estimating the SFPM, the TE was predicted using the formula given in Eq. 5.3, and TE score was regressed against socioeconomic variables (Eq. 5.4) to find their impact on TE.

$$\begin{aligned} \mathrm{TE}_i = \delta_0 &+ \delta_1 \ln \text{age of HHH} + \delta_2 \ln \text{education of HHH} \\ &+ \delta_3 \ln \text{family labor} + \delta_4 \ln \text{off} - \text{farm income} \\ &+ \delta_5 \ln \text{irrigation cost} + \delta_6 \ln \text{land rent} + \delta_7 \ln \text{experience} \\ &+ \delta_8 \,\text{training} + \delta_9 \,\text{Chitwan} + \omega_i \end{aligned} \tag{5.4}$$

Data for family labor were calculated using the active labor force unit (LFU).[1] Similarly, the household's livestock possession was estimated through the livestock standard unit (LSU).[2] The LSU represents the amount of organic manure used by households in the crop field because there is no system for trading organic manure in the study area, and whatever manure is produced at the household is incorporated in the crop field. Irrigation cost was estimated considering electricity and labor costs in case of tube well use, whereas it is labor cost for those using an irrigation canal. Similarly, land rent is the value of share crop land estimated by household.

Here, δ represents the parameters associated with socioeconomic variables, and ω_i is the error term. The sign of socioeconomic variables and their description are presented in Table 5.1. Of these variables, the impact of age and education of HHH were hypothesized to be positive (Ali and Flinn 1989; Rahman 2003). Training and experience in seed production are capacity enhancement variables, and these variables were assumed to have a positive influence on TE (Rahman 2003). Similarly, the influence of irrigation cost on TE was assumed as negative. The influence of land rent was hypothesized as positive because this represents the land quality and has a direct link with crop yield (Ghaderzadeh and Rahimi 2008). The majority of the labor force in rural areas is supplied by the farmers' family members, which makes it easy to access labor when required, and easy accessibility of labor might contribute positively to production. So, family labor at a household was hypothesized to have a positive influence on TE. Similarly, household income from an off-farm source was also assumed to have a positive contribution to TE, as farmers

[1] LFU is the measurement of labor force, wherein persons from 15 to 59 years old, regardless of their sex, were categorized as 1 person = 1 LFU, but for children (10–14 years old) and elderly persons (>59 years old), 1 person = 0.5 LFU.

[2] LSU is the aggregate of different types of livestock kept at a household in a standard unit calculated using the following equivalents: 1 adult buffalo = 1 LSU, 1 immature buffalo = 0.5 LSU, 1 cow = 0.8 LSU, 1 calf = 0.4 LSU, 1 pig = 0.3 LSU, 1 sheep or goat = 0.2 LSU, and 1 poultry or pigeon = 0.1 LSU (Khanal and Maharjan 2014).

Table 5.1 Socioeconomic variables used in the model

Variable	Description	Coefficient	Expected sign
Age of head of household (HHH)	Age of HHH in years	δ_1	+/−
Education of HHH	Formal schooling years of HHH in years	δ_2	+/−
Training	If household attended agricultural training, = 1, and 0 for otherwise	δ_3	+
Labor	Labor force unit (LFU) available at households	δ_4	+
Irrigation cost	NRs ha^{-1} in the cropping season	δ_5	−
Land rent	NRs ha^{-1} per cropping season	δ_6	+
Experience	Years of household participating in seed marketing	δ_7	+
Off-farm income	Amount of cash money households receive from members in a year (NRs $year^{-1}$)	δ_8	+
Chitwan	If household from Chitwan = 1, and 0 for otherwise	δ_9	+

could accomplish agricultural activities in timely fashion if they have access to off-farm resources (Ali and Flinn 1989; Wang et al. 1996; Rahman 2003).

We have used Chitwan as the dummy variable in the analysis, considering that this district might have a positive influence on TE for reasons of better infrastructure and extension facility. Before running the stochastic frontier production model and ordinary least squares model, data were validated for multicollinearity, heteroscedasticity, and endogeneity issues.

5.3 Results and Discussion

5.3.1 Summary of Study Variables

The average age of a head of household (HHH) is 46.8 years, but this varies from 16 to 78 years. In general, HHHs have lower secondary school level education (7.96 years), but this varies, from 0 to 18 years. Not having formal schooling does not explicitly indicate no education, because many household heads are also engaged in informal education schemes supported by nongovernment organizations (NGOs); however, this study has not captured this factor in the analysis. Most of the households (78 %) receive training from government agencies and NGOs, and this training represents general orientation about crop production including site selection, crop husbandry, and plant protection.

As rice is a water-loving crop, irrigation frequency is substantially higher (three to four times in a month) than that of wheat (two to three times in a growing season). However, there is not much difference in irrigation costs between rice and wheat, indicating that irrigation costs during the wheat growing season (winter

Table 5.2 Summary of the explanatory variables

Variable inputs	Overall	Chitwan	Siraha	Kailali
Age of HHH	46.8 ± 11.43	49.8 ± 12.05	46.1 ± 10.9	44.6 ± 10.8
Education of HHH	7.96 ± 4.02	10.6 ± 2.97	6.8 ± 4.1	6.4 ± 3.4
Training	0.78 ± 0.41	0.81 ± 0.39	0.68 ± 0.46	0.85 ± 0.36
Family labor	3.4 ± 0.37	2.21 ± 1.35	4.12 ± 1.25	4.23 ± 2.04
Irrigation cost (rice)	1,863 ± 1,135	1,822 ± 1,369	2,267 ± 1,169	1,500 ± 605
Irrigation cost (wheat)	1,545 ± 945	1,224 ± 654	1,854 ± 876	1,721 ± 876
Land rent	6,145 ± 1,827	8,310 ± 872	4,125 ± 655	6,000 ± 2,540
Experience	4.37 ± 0.97	5.8 ± 0.35	4.2 ± 0.57	4.1 ± 0.52
Off-farm income	42,998 ± 52,622	53,990 ± 14,566	43,510 ± 16,540	25,950 ± 16,452

Source: Survey 2011
SD standard deviation, 1 US$ = NRs. 82.96

season) is substantially higher than those in the rice season (rainy season). Land rent varies across households and the district, and it has been used as a proxy variable to represent land quality and was measured as value of share cropland.

There is also quite a variation in land rent, ranging from NRs 3,000 to NRs 9,000 ha^{-1} per cropping season (i.e., 6 months) in the study area, and this figure is also higher in Chitwan as compared to the other two districts (Table 5.2).

5.3.2 *Findings from the Stochastic Frontier Production Model*

Table 5.3 presents the findings from the stochastic frontier production model. The significant log-likelihood test (Wald test) signifies that variables chosen fit the model well. Moreover, the likelihood ratio test for the "absence of inefficiency in the model" is rejected, which indicates that the inefficiency effect explained in the model is higher than random noise.

The marginal effect of input variables was also estimated because the stochastic frontier model was run through the maximum likelihood method, and coefficients of input variables do not represent their average impact on the dependent variable. All the input variables except labor and land have a positive response on yield. The marginal effect of labor is −0.11, indicating that 1 % increase in LFU leads to decrease in rice yield by 0.11 %. The impact of chemical fertilizer and livestock is also positive.

Similarly, all the input variables except livestock show a significant positive impact on wheat yield. The marginal impact of seed on wheat yield is 0.38, which means wheat yield could be increased by 0.38 % with a 1 % increase in seed rate. This result shows that farmers use seed at a rate of 115.69 kg ha^{-1}, which is less

Table 5.3 Effect of explanatory variables on crop yield

Variables	Rice			Wheat		
	Coefficient	p value	Marginal effect	Coefficient	p value	Marginal effect
Seed	0.335	0.75	0.158	0.28	0.001	0.125
Labor	−0.221	0.017**	−0.111	−0.154	0.037	−0.128
Chemical fertilizer	0.089	0.081*	0.089	0.315	0.002	0.235
Livestock	0.032	0.057*	0.020	0.012	0.549	0.009
Operational land	−0.018	1.90	−0.019	0.053	0.041	0.023
Constant	7.40	0.124		2.01	0.189	
Model characteristics	Log likelihood: −110.38**, $\sigma^2 = 0.143$, $\lambda = 2.35$, Likelihood ratio = 46.58***, $n = 121$			Log likelihood: −110.38**, $\sigma^2 = 0.143$, $\lambda = 2.35$, Likelihood ratio = 46.58***, $n = 121$		

*,**,*** indicate significance at 10 %, 5 %, and 1 % level, respectively

than the recommendation made in wheat production (120 kg ha^{-1}) under irrigated condition in the Tarai region of Nepal (MoAC 2010). The marginal effect of labor is −0.154, which is significant at the less than 5 % level of significance, implying that 1 % increase in labor leads to decrease in wheat yield by 0.154 %. The reason behind the negative response of labor to wheat yield might be that most of the labor involved in seed production is supplied by family members and they are unpaid, and in the absence of better job opportunities in the rural areas they could spend more of their time than is required in farming. Sohail et al. (2012) have also found similar results in Pakistan.

5.3.3 *Farmers' Technical Efficiency*

The results show that the efficiency of farmers in rice seed production is 81 %: it varies from 36.7 to 95.2 %, suggesting that farmers could improve their efficiency in rice seed production by 19 %. Efficiency level also differs among the districts. The average efficiency of farmers in Chitwan, Siraha, and Kailali is 85.1 %, 75.8 %, and 81.8 %, respectively. Previous studies have also identified a wide range of efficiency among the farmers. For example, Kyi and Oppen (1999) found average efficiency of farmers to be 88 %, ranging from 39 % to 93 % in irrigated rice production in Myanmar. Similarly, Idiong (2007) found the efficiency of Nigerian rice farmers to be 77 %, ranging from 48 % to 99 %. The recent study by Piya et al. (2012) found the efficiency of Nepalese rice growers to be 74 %, ranging from 35 to 100 %. All these studies also indicated wide variation in efficiency level among the study households.

Farmers are 78.3 % efficient in seed production of wheat. However, production efficiency varies from 38.6 to 94.6 %, so that farmers could improve wheat production efficiency by 21.7 % (range, 5.4–61.4 %). Moreover, more than 70 % households are above 70 % efficiency level in seed production. Studies from other developing countries have also shown a wide range of efficiency among wheat growers. For example, Hassan and Ahmad (2005) reported 93.6 % (range, 58–98.5 %) efficiency in the mixed farming system of Pakistan. Similarly, Ghaderzadeh and Rahimi (2008) found the technical efficiency of farmers to be 65.6 % (range, 30–94 %) for rain-fed wheat farming in Iran. Similarly, Sohail et al. (2012) estimated 60 % technical efficiency (range, 25−85 %) of wheat growers in Pakistan.

5.3.4 Impact of Socioeconomic Variables on Technical Efficiency

A total of nine socioeconomic variables were tested for their impact on TE, and the results show that direction of impact of most of the variables is as per the expected sign, except for family labor, training, and age of HHH (Table 5.4).

In rice, education level of the HHH, household experience on seed marketing, land rent, and location have a significant positive influence on TE. The significant positive response of education of HHH on TE indicates that efficiency of the farmers would be further increased with the current education level (7.8 years).

In the study area, 35.6 % of HHH are still below the primary education level (1–5 years of formal schooling), and this category of households experiences less yield (16.6 %), higher yield loss (17.6 %), and operates in the lower efficiency level

Table 5.4 Influence of socioeconomic variables on technical efficiency

	Rice	Wheat
Variables	Coefficient (*p* value)	Coefficient (*p* value)
Age of HHH	−0.0325 (0.251)	0.031 (0.29)
Education of HHH	0.0042 (0.06)*	0.023 (0.416)
Family labor	−0.014 (0.876)	−0.016 (0.392)
Training	0.052 (0.264)	0.010 (0.680)
Irrigation cost	0.008 (0.968)	0.03 (0.02)**
Land rent	0.045 (0.005)***	0.152 (0.0001)***
Experience	0.063 (0.0112)**	0.059 (0.31)
Off-farm income	0.0545 (0.0683)	0.008 (0.297)
Chitwan	0.034 (0.13)	0.023 (0.387)
Constant	0.023 (0.04)**	−0.318 (0.127)
Goodness of fit	R^2, 0.56; adjusted R^2, 0.52	R^2, 0.46, adjusted R^2, 0.43

*, **,*** indicate significance at 10 % level, 5 % level, and 1 % level, respectively. Values given in the parentheses are the *p* values, $n = 121$

Table 5.5 Observed yield, yield loss, and technical efficiency of significant variables

Variables	n	Observed yield	Yield loss[a]	Technical efficiency (%)
Education of HHH				
Primary (≤5 years)	36	3,453	1,203	0.73
Above (>5 years)	85	4,027	991	0.79
p value		0.002***	0.005***	0.001***
Land rent (NRs for 6 months)				
≤7,000	50	3,655	1,108	0.758
>7,000	71	4,571	873	0.831
p value		0.001***	0.008***	0.001***
Experience (years)				
≤3	45	3,648	1,263	0.7568
>3	76	3,904	1,016	0.783
p value		0.006***	0.0001***	0.0008***

*,**,*** indicate difference between categories by 10 %, 5 %, and 1 %, respectively
[a]Yield loss = maximum possible production − observed production, and maximum possible production = observed production/TE

(8.7 %) compared to those having higher education. Previous studies have also shown the positive response of education on TE of farmers (Idiong 2007; Piya et al. 2012). The better performance of higher-educated HHH might reflect their better analytical capability and extension contact (Battese and Coelli 1995). The study also shows that a 1 % increase in land rent leads to a 0.045 % increase in TE. Land rent represents land quality, and to compare the level of land quality with efficiency, households are divided into good land quality households (>NRs 7,000/season/ha) and poor land quality households (<NRs 7,000/season/ha). It was found that the households in the former category achieved higher yield (20 %), experienced less production loss (21 %), and these farmers operate at a higher TE level (8.78 %) as compared to their counterparts (Table 5.5).

The study also shows the significant positive influence of household experience with seed production on TE. As shown in Table 5.4, 1 % increment in years of experience leads to increased efficiency of farmers by 0.063 %. Households with more than 3 years of experience in seed production realized 3.6 % higher efficiency as compared to those having less experience ($p = 0.047$). Other studies have also shown a positive link between farmer experience and level of technical efficiency (Kyi and Oppen 1999; Idiong 2007). The reason behind the higher efficiency of experienced farmers might be their better skills in managing resources compared to less experienced farmers.

As for rice, land rent has significant positive impact on TE in wheat production. As shown in Table 5.4, the coefficient for the impact of land rent on TE is 0.152, which means 1 % increase in land rent leads to an increase of TE in wheat seed production by 15.2 %. Thus, households evaluating higher quality for their land tend to be more efficient than those evaluating their land as lower quality. As shown in Table 5.6, households evaluating their land as being of better quality obtained a

Table 5.6 Comparison of yield loss and technical efficiency among different categories

Variables	*n*	Observed yield (kg ha^{-1})	Yield loss (kg ha^{-1})	Technical efficiency (%)
Irrigation cost (NRs ha^{-1})				
≤1,000	44	2,833	635	81.22
>1,000	77	2,719	726	77.10
p value		0.03**	0.08*	0.04**
Land rent (NRs ha^{-1})				
≤7,000	143	2,504	733	79.72
>7,000	37	2,806	595	83.050
p value		0.05**	0.013**	0.024**

*, ** indicate significance at 10 % and 5 %, respectively

12.06 % higher yield, experienced 18.8 % less yield loss, and operated at a 4.17 % higher efficiency level than their counterparts. The positive response of land quality on efficiency of wheat production was also reported by previous studies (Sharma et al. 1999; Kamruzzaman and Islam 2008). In a focus group discussion, farmers stated that soil fertility and access to a irrigation facility are the major determining factors for land quality by farmers.

Another socioeconomic variable having significant impact on TE is irrigation cost. One unit increase in irrigation cost leads to increasing TE by 3 % (Table 5.4). There is also a significant difference between households paying higher (>NRs 1,000) and lower (≤1,000) irrigation cost in observed yield, yield loss, and TE (Table 5.6).

Irrigation cost in this case mainly represents labor cost involved in repair and maintenance of the water canal as only 3.5 % of households used a tube well for irrigating wheat fields. Better performance of households realizing higher irrigation costs means their better access to an irrigation facility because households tend to pay higher costs where water is available in irrigation canals during winter season. Previous studies have also shown the positive influence of irrigation on farmer TE (Wang et al. 1996; Sharma et al. 1999). Also, households with better access to an irrigation facility could motivate the farmers to combine other inputs (such as chemical fertilizer) for higher production (Wang et al. 1996; Ghaderzadeh and Rahimi 2008).

5.4 Conclusion

This chapter has analyzed the technical efficiency of seed-producing households in rice and wheat seed production using the most commonly used inputs through a stochastic frontier production model. The efficiency level also represents the potential benefits households realized in the seed production stage by utilizing existing resources. Results indicate a wide variation in efficiency level among the seed producers, and the variation is mainly explained by the HHH's education level, land rent, irrigation cost, and experience. From these findings, three policy measures could be recommended. The education of household heads is low in the study area, and it might not be feasible to provide formal education, considering their age, but the provision of practical oriented/field-based training would contribute to their efficiency. The second policy measure would be the provision of an irrigation facility in a seed production area. The third policy recommendation from this study is land management, suggesting the adoption of soil fertility management measures to improve the efficiency of farmers in rice and wheat seed production.

References

Aigner D, Lovell CAK, Schmidt P (1977) Formulation and estimation of stochastic frontier production model. J Econ 6:21–37

Ali M, Flinn J (1989) Profit efficiency among Basmati rice producers in Pakistan Punjab. Am J Agric Econ 71(2):303–310

Battese GE, Coelli TJ (1995) A model for technical inefficiency effects in a stochastic frontier production function for panel data. Empir Econ 20:325–332

Dung KT, Sunalde ZM, Pade VO, Mckinley JD, Gercia YT, Bello AL (2011) Technical efficiency of resource conserving technologies in rice-wheat system: the case of Bihar and Eastern Uttar Pradesh in India. Agric Econ Res Rev 24:201–210

Farrell MJ (1957) The measurement of productivity efficiency. J R Stat Soc 120:253–290

Ghaderzadeh H, Rahimi MH (2008) Estimation of technical efficiency of wheat farm: a case study in Kurdistan Province, Iran. Am-Euras J Agric Environ Sci 4(1):104–109

Hassan S, Ahmad B (2005) Technical efficiency of wheat farmers in mixed farming system of Punjab, Pakistan. Int J Agric Biol 7(3). Retrieved March 10, 2012, from Ijab.org

Idiong IC (2007) Estimation of farmer level technical efficiency in small scale swamp rice production in cross river state, Nigeria: a stochastic frontier approach. World Agric Sci 4:45–60

Kalirajan KP (1999) The importance of effective and efficient use in the adoption of technologies: a micro panel data analysis. J Prod Anal 2:113–126

Khanal NP, Maharjan K (2014) Adoption of improved rice varieties in the Tarai region of Nepal. In: Maharjan KL (ed) Communities and livelihood strategies in developing countries. Springer, Tokyo, pp 49–64. doi:10.1007/798-4-431-54774-7-4

Kamruzzaman M, Islam MH (2008) Technical efficiency in wheat growers in some selected sites of Dinajpur district of Bangladesh. Bangladesh J Agric Res 33(3):363–373

Kyi T, Oppen MV (1999) Stochastic frontier production function and technical efficiency estimation: a case study on irrigated rice in Myanmar. Retrieved 12 Sept 2012, from ftp://ftp3.gwdg.de/pub/tropentag/proceedings/1999/.../STD_C6.pdf

Meeusen W, van den Broeck J (1977) Efficiency estimates from Cobb-Douglas production function with composed error. Int Econ Rev 18:435–444

MoAC (2010) Statistical information on Nepalese agriculture. Ministry of Agriculture and Cooperatives, Kathmandu, Nepal

Morris ML, Rusike J, Smale M (1998) Maize seed industries: a conceptual framework. In: Morris ML (ed) Maize seed industries in developing countries. Lynne Rienner, Boulder

Piya S, Kimanami A, Yagi H (2012) Comparing the technical efficiency of rice farmers in urban and rural areas: a case study from Nepal. Trends Agric Econ 5(2):48–60

Rahman S (2003) Profit efficiency among Bangladeshi rice farmers. Food Policy 28:487–503. Retrieved February 25, 2012, from http://www.sciencedirect.com/science/article/pii/S0306919203000745

Sharma KR, Leung P, Zaleski HM (1999) Technical, allocative and economic efficiency in swine production in Hawaii: a comparison of parametric and non-parametric approaches. Agric Econ 20:23–25

Sohail N, Latif K, Abbas N, Shahid M (2012) Estimation of technical efficiency and investigation of efficiency variables in wheat production: a case of District Sargodha, Pakistan

Schultz TW (1964) Transforming traditional agriculture. Yale University Press, New Haven

Wang J, Cramer GL, Wailes EJ (1996) Production efficiency of Chinese agriculture: evidence from rural household survey data. Agric Econ 15:17–28

World Bank (2011) World development indicators. Retrieved 28 May 2012, from http://data.worldbank.org/indicator/SP.RUR.TOTL.ZS

Chapter 6
Profit Efficiency in Seed Production Under Rice–Wheat Farming

Abstract Profit efficiency is a tool to measure the performance of economic agents in utilizing their resources. This chapter measures profit efficiency of seed producers in using chemical fertilizer, labor, animal manure and operational land in rice-wheat cropping system using stochastic frontier profit function. Data for this study was collected from 180 households involved in rice and wheat seed production across the three Tarai districts: Siraha, Chitwan, and Kailali of Nepal. Result shows that households are 41.1 % efficient in utilizing the aforementioned inputs, and the efficiency level among the households varies from 6 to 84 %. This variation is mainly explained by gender, physical structure, technical empowerment, incentive system, and diversification. This study recommends the integration of social benefit and technical empowerment indices along with economic benefit while evaluating the performance of community seed production.

Keywords Profit efficiency • Rice-wheat system • Community seed production • Nepal

6.1 Introduction

The importance of community seed production (CSP) in improving farmers' access to diversified varieties of open-pollinated food crops in a cost-effective way has been recognized in the recent literature (David 2004; Witcombe et al. 2010; Khanal and Maharjan 2010; Srinivas et al. 2010). The assumption of cost-effectiveness is mainly based on farmers' potentials in reducing costs for seed production and distribution as these activities are implemented at local levels (narrow geographic boundary). In addition, CSP provides ample opportunities to promote local landraces and famer-led technological innovations. Despite these potential benefits of CSP, very limited studies have been published on farmer performance in this system (Poudel et al. 2003; Srinivas et al. 2010; Witcombe et al. 2010). The measurement of efficiency in utilizing resources is the popular approach in understanding farmers' performance in agriculture, including seed production. The efficiency can be measured from production and profit functions; farmers' technical efficiency in utilizing resources has been already discussed in Chap. 5. This chapter analyzes the efficiency of farmers in utilizing commonly used household-level

N.P. Khanal, K.L. Maharjan, *Community Seed Production Sustainability in Rice–Wheat Farming*, DOI 10.1007/978-4-431-55474-5_6

resources from profit function to measure profit efficiency (PE). This approach is considered more appropriate when farmers face variation in output price (Ali and Flinn 1989). The PE combines both production and cost functions, as any deviation in allocation decision is translated into profit function, and this it supplements the research findings discussed in Chap. 5.

There are few studies on PE (Rahman 2003; Wang et al. 1996; Alli and Flinn 1989), and these studies have identified technical, communication, infrastructure, and credit constraints for increasing PE in agricultural crops. These studies have considered a single crop in the analysis; however, consideration of system perspective would be necessary to make practicable recommendation to farmers because farmers apply soil conservation practices including animal manure in their fields considering the existing cropping system (Khanal and Maharjan 2014). Therefore, this study intends to measure PE of Nepalese seed producers under the rice-wheat system from a cropping system perspective.

6.2 Methodology

6.2.1 Study Area and Sampling Technique

Data from this study were collected from households involved in the CSP scheme under the rice-wheat (R-W) system. Three districts, Siraha, Chitwan, and Kailali, were purposively selected to make the study representative of the Tarai region (plains area with altitudinal range from 70 to 650 m above sea level). From each of these districts, four farmers' seed producer organizations (SPOs) with at least 2 years of experience in rice and wheat seed production were selected in consultation with District Agriculture Development Offices. Then, 15 seed-producing households from each of the selected organizations were randomly selected to make the sample size of 180. From the selected households, costs of production and profits from rice and wheat seed production were measured.

6.2.2 Empirical Model

General Background of the Model

This study utilizes the stochastic frontier profit (SFP) model proposed by Battesse and Coelli (1995) for measuring PE. In this model, PE is defined as the ratio of observed profit and frontier profit. There are two approaches in PE measurement under the SFP model: (1) simultaneous modeling of major explanatory variables associated with the profit relation, and those creating inefficiency, and (2) adoption of the two-stage procedure. In the latter procedure, the SFP model is run using the major variables involved in the profit relation, and PE score is estimated. Then, possible inefficiency causing variables are regressed against the efficiency score. The first approach is more efficient than the second one because the second approach assumes the independence of error terms in first and second stages (Battesse and Coelli 1995; Rahman 2003). Theoretically, profit function is given in Eq. 6.1:

$$\pi_i = f\left(p_{ij}, z_{ik}\right).\exp(\mu_i) \tag{6.1}$$

where π_i is the normalized profit of the ith household, which is defined as the difference between gross revenue less than total variable costs. p_{ij} is the price of jth variable inputs faced by ith farmer divided by output price. Similarly, z_{ik} is the level of k fixed factors in the ith household. Similarly, μ_i is the error term representing inefficiency effect in the model and is considered to make truncation 0 of the normal distribution with mean $\mu_i = \delta_0 + \Sigma\delta_d w_d$ and variance σ_μ^2. v_i is the two-sided random error term, which represents the variation occurred in the model resulting from random factors including measurement errors, disease, and climate factors. w represents the factors associated with inefficiency in the model, d is the types of inefficiency factors, and δ is the vector of coefficients in the inefficiency model. The profit efficiency of farmers in the context of SFP model is measured using Eq. 6.2.

$$\mathrm{PE_i} = \mathrm{E[Exp} - (\mu_i)|v_i] = E\left[\mathrm{Exp} - \delta - \sum \delta w|v_i\right] \tag{6.2}$$

The variance parameters of the model are expressed in terms of $\sigma^2 = \sigma_\mu^2 + \sigma_v^2$, and $\gamma = \sigma_\mu^2/\sigma^2$. Here, γ explains the proportion of inefficiency effect explained in the model, and its values range from 0 to 1.

Specification of the Model and Variables

This study adopts the SFP model in trans log functional form for data analysis (Eq. 6.3). The decision to use this functional form was based on the significant likelihood ratio test between Cobb–Douglas and trans log forms ($p = 0.024$); this shows the rationale for considering the interaction effect among variables in the model, the function form of which is presented in Eq. 6.4:

$$\begin{aligned}\pi' &= \alpha_0 + \sum \ln \mathrm{p}'_j + \frac{1}{2}\sum\sum \tau_{jk} \ln \mathrm{p}_{jk} \ln \mathrm{p}'_k + \sum\sum \text{Ø}_{jl} \ln x'_j \ln z_l \\ &+ \sum \beta_l \ln z_l + \frac{1}{2}\sum\sum \varphi_{lt} \ln z_l \ln z_t + v - \mu\end{aligned} \tag{6.3}$$

$$\mu = \delta_0 + \sum \delta_d W_d + \omega \tag{6.4}$$

where

π' = restricted profit (total revenue minus total variable cost) normalized by output price (p_y)

p_j' = price of the jth inputs (p_j) normalized by p_y; J = (1) average labor wage, (2) fertilizer price; z_l = quantity of fixed inputs: (1) livestock, and (2) operational land for seed production (ha).

ln is natural log; and α, τ, $\varnothing$, φ, and δ are the parameters to be estimated. Similarly, k and t are the multiplication factors. The variables representing inefficiency effect (w) include (1) gender (this is a dummy variable, and a male-headed household is expected to have positive influence on PE); (2) age of household head (HHH), which represents farmer's experience in crop husbandry, and (3) education of HHH, that indicates innovativeness and rationale decision in input use, both of which were assumed to have a positive impact on PE (Table 6.1). Another demographic variable (4) dependency ratio was assumed to have a negative effect on PE considering shortage/risk of labor supply in seed production activities. The other variables expected to have positive impact on PE are (5) physical index, (6) incentive index, (7) technical index, and (8) soil health index.

The physical index represents the households' physical assets such as house type, irrigation structure, and availability of a tractor at a household. House type affects the seed price as concrete-roofed households tend to get a higher price for their seed because of their ability to protect seeds from pests, diseases, and excess moisture. Similarly, access of irrigation structure (canal) and tractors enhances labor use efficiency and thereby the PE. The variables associated with the incentive index represent a household's realization of economic and social benefits being involved in seed production, and the consideration of these benefits is crucial in CSP, especially those operating in the initial phase of their development (Khanal and Maharjan 2014). Being engaged in CSP, households receive various training, visits, and orientation from development agencies. These activities empower the farmers in seed production and marketing, and the outcome of these activities is reflected in profit efficiency. Similarly, soil conservation practices contribute to the soil organic matter, and the impact of these practices is depicted in PE. Index values for infrastructure, incentive, technical empowerment, and soil health were estimated separately using the principal component analysis (PCA)[1] procedure. The detail of the components under each index is given in Table 6.2. Similarly, the R-W system was assumed diversified if farmers grew any additional crops after wheat harvest, and its impact on PE was assumed positive considering the contribution of soil health to PE (Khanal et al. 2008).

[1] PCA = $\alpha_1 x_1 + \alpha_2 x_2 + \cdots, \alpha_m x_n$, where PC = the principal component, α_1 is weight of variable 1 (value ranging from 1 to n), and $x_1 \ldots x_n$ are the observations.

Table 6.1 Variables and their expected relationship with profitability and profit inefficiency

Variables	Definition	Mean ± SD	Expected sign
Profitability			
Output price	Average price of seed (NRs)	28.0 ± 7.05	
Profit	Profit (NRs)/ha	37,447 ± 24,163	
Labor (L)	Labor cost (NRs/day)	257.8 ± 86.05	+
Fertilizer (F)	Average price (NRs/kg)	32.54 ± 6.33	+
Livestock (Li)	LSU[a]/household	3.86 ± 2.97	+
Land (La)	Operational land (ha)	0.86 ± 0.26	+
Profit inefficiency			
Gender of household head (HHH)	1 if male, and 0 otherwise	0.744 ± 0.44	–
Age of HHH	Years	44.5 ± 1.50	–
Education of HHH	Education of HHH (years)	7.96 ± 4.02	–
Dependency ratio[b]	Proportion of dependents in HHH	0.441 ± 0.19	+
Infrastructural index	As explained in Table 6.2	1.02 ± 0.58	–
Incentive index	As explained in Table 6.2	1.847 ± 0.67	–
Technical empowerment	As explained in Table 6.2	1.433 ± 0.60	–
Soil health index	As explained in Table 6.2	0.915 ± 0.61	–
Diversification index	1 = if farmers grow additional crops than rice and wheat, 0 = otherwise	0.192 ± 0.24	–

I US$ = NRs. 82.96

[a]LSU is the aggregates of different types of livestock kept at the household in standard unit calculated using the following equivalents: 1 adult buffalo = 1 LSU, I immature buffalo = 0.5 LSU, 1 cow = 0.8 LSU, 1 calf = 0.4 LSU, 1 pig = 0.3 LSU, 1 sheep or goat = 0.2 LSU, and 1 poultry or pigeon = 0.1 LSU

[b]Dependency ratio (DR) was calculated using the following formula: DR = number of inactive members divided by total members in the household. People outside the age range of 15–59 years and handicapped people were considered dependent (Khanal and Maharjan 2013)

6.2.3 *Description of Variables*

Farmers grow rice (0.873 ha) and wheat (0.473 ha) for seed, and the proportion of the total operational land occupied by these crops in the respective crop production season (rice, June to September; wheat, October to March) is 68.9 % and 37.36 %, respectively. Seed, labor, and chemical fertilizers are the major variable inputs, and they share 5 %, 70 %, and 25 % of the total variable costs. Chemical fertilizers applied in the R-W system are in the form of diammonium phosphate (containing 18 % nitrogen and 46 % phosphorus), muriate of potash (containing 60 %

Table 6.2 Summary of different indices created to use as explanatory variables in model

Types	Specification	PCA weight
(A) **Infrastructural index**: variability explained in component one = 57.04 %		
1. House type	1 if concrete roof (33.8 %), and 0 otherwise	0.5231
2. Irrigation facility	1 if access to irrigation canal (55 %), and 0 otherwise	0.5875
3. Tractor use	1 if use tractor (84.4 %), and 0 otherwise	0.6174
(B) **Incentive index**: variability explained in component one = 50.28 %		
1. Price	1 if higher price for seed than grain (85.5 %), and 0 otherwise	0.4871
2. Training	1 if received training from/through SPOs (78.3 %), and 0 otherwise	0.4107
3. Loan	1 if taken loan from SPO (43.8 %), and 0 otherwise	0.2735
4. Source seed	1 if source seed received through SPOs (92.2 %), and 0 otherwise	0.4627
5. Prestige	1 if realized social prestige (78.8 %), and 0 otherwise	0.4424
6. Profit in seed processing	1 if get economic profit generated by SPOs in marketing (64.8 %), and 0 otherwise	0.3310
(C) **Technical empowerment index:** variability explained in component one = 53.2 %		
1. Knowledge of source seed	1 if heard any one of the seed categories – breeder/ foundation (77.22 %), and 0 otherwise	0.4504
2. Households' engagement in rogueing	1 if household removes unwanted/diseased plants (86.67 %), 0 otherwise	0.530
3. SPO's facilitation in rogueing	1 if SPO's members facilitate in rogueing (75 %), and 0 otherwise	0.569
4. Seed lab's inspection in field	1 if field inspection by seed lab (46.6 %), and 0 otherwise	0.437
(D) **Soil health index**: variability explained in component one = 33.4 %		
1. Zero tillage (ZT)	1 if ZT adopted (46.6 %), and 0 otherwise	0.5188
2. Compost improvement	1 if used effective microorganisms (20 %), and 0 others	0.3560
3. Green biomass (GB)	1 if GB incorporated (51.6 %), and 0 otherwise	0.3233
4. Seed priming (SP)	1 if SP adopted (60.5 %), and 0 otherwise	0.4726
5. System of rice intensification (SRI)	1 if SRI adopted (12.22 %), and 0 otherwise	0.3728
6. Integrated pest management (IPM)	1 if botanical pesticides adopted (27.7 %), and 0 otherwise	0.3705

Figures in the parentheses indicate adoption

phosphorus), and urea (containing 46 % nitrogen). These fertilizers are applied both in rice and in wheat at the rate of 91.6 kg nitrogen, 60.5 kg phosphorus, and 31.2 kg potash, and rice shares 60 % of the total chemical fertilizer consumption. Livestock manure is the major source of organic matter in seed production under the R-W system, and the manure is applied targeting rice. On average, farmers make a profit at the rate of NRs 37,447 ha^{-1} (Table 6.1) from seed production of these crops, and it varies from NRs 12,616 ha^{-1} to 270,925 ha^{-1}. In the process of profit estimation, gross revenue from seed as well as by-product (straw) was considered. The per unit labor wage was calculated by taking average of male and female labor wage, and chemical fertilizers and output price estimation were calculated similarly. As indicated in Table 6.2, only 33.8 % of households have a concrete roof, less than half of them use a canal irrigation facility to irrigate crops, and bullocks are the major source of animal power. It was also found that households realize incentives from various sources while engaging in seed production; however, farmers perceive that access to source seed (92.2 %), output price (85.5 %), social prestige (78.8 %), training (78.3 %), profit (64.8 %), and easier access to loans (43.8) are the major incentives in order of importance. Moreover, seed producers realize benefits from technical empowerment as well as soil health improvement practices.

6.2.4 *Results*

This study shows that the variables chosen for this study fit the model well. The significant log likelihood ($p < 0.000$) indicates that coefficients of explanatory variables are significantly different from zero. Similarly, the likelihood ratio test for $\mu = 0$ is rejected at $p < 0.000$. Similarly, the value of γ is 0.677, which means that 67.7 % of the inefficiency effect is explained in the model (Table 6.3). All these figures indicate the goodness of fit of the model. Labor wage and fertilizer price show positive impact on profitability of the R-W system, whereas the impact of livestock and operational land on the profitability is negative. There is also significant positive influence of fertilizer and land size interaction on profitability of the R-W system.

Overall, households are 41.1 % efficient in utilizing the given resources (labor, chemical fertilizer, animal manure, and land), but there remains a 58.9 % inefficiency effect in the allocation of resources. Moreover, the efficiency level varies from 6 to 88 %. Efficiency level also varies across the district. Farmers of Chitwan are more efficient (46.9 %) than those of Kailali (45.8 %) and Sirha (31.5 %) districts (Fig. 6.1). Previous studies have also shown variation in efficiency levels among the households in agricultural production. For example, Rahman (2003) found 77 % PE with range from 16.8 to 94.8 % in Basmati rice production among Bangladeshi farmers. Ali and Flinn (1989) reported mean PE of 69 % with a range of 13–95 % for Basmati rice producers in Punjab, Pakistan. Similarly, Assa et al. (2012) found 74 % (range, 31–99 %) efficiency among Malawi's potato farmers.

Table 6.3 Output from stochastic profit frontier model

Variables	Coefficients	Standard deviation	*p* value
Labor wage (L)	2.493	3.118	0.424
Fertilizer price (F)	7.734	3.313	0.014**
Livestock unit (Li)	−0.175	0.335	0.602
Operational land (La)	−0.304	0.586	0.604
0.5*(L*L)	−1.546	1.382	0.263
0.5*(F*F)	−3.586	1.924	0.062*
0.5*(Li*Li)	0.177	0.090	0.048**
0.5*(La*La)	0.049	0.123	0.690
L*F	−1.506	1.213	0.214
L*Li	−0.235	0.211	0.124
L*La	−0.840	0.114	0.463
F*Li	0.464	0.408	0.255
F*La	−0.807	0.460	0.079**
Li*La	0.524	0.324	0.215
Constant	4.522	3.539	0.208
Explanatory variables			
Gender	−0.663	0.329	0.044**
Age of HHH	0.003	0.013	0.771
Education of HHH	−0.0329	0.064	0.610
Dependency ratio	−0.718	0.691	0.299
Physical structure	−0.166	0.46	0.097*
Incentive index	−0.077	0.016	0.037**
Technical empowerment	−0.246	0.151	0.012**
Soil health index	−0.041	0.108	0.425
Diversification index	−0.028	0.021	0.027**
Constant	1.972	0.998	0.158

$n = 148$, Wald chi-square 2 (13) = 124;1 ($p = 0.000$); likelihood ratio test (for sigma $\mu = 0$) = 18.76 ($p = 0.000$). $\sigma^2 = 0.425$, $\sigma_v^2 = 0.2025$, $\gamma = 0.677$

*,**, *** indicate significant at 10 %, 5 %, and 1 % level of significance, respectively

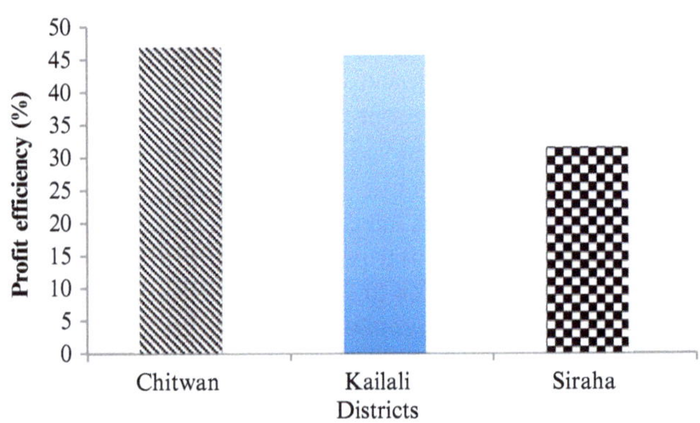

Fig. 6.1 Profit efficiency of farmers across the districts (From Survey 2011)

The variation in efficiency level could be explained by understanding their relationship with inefficiency-inducing variables. All the inefficiency variables included in the model except age of HHH showed their direction of impact as per expectations. However, significant influence was found in gender of HHH, infrastructural index, incentive index, technical empowerment index, and diversification index. Observed profit, profit loss[2], and PE were computed and compared between categories of the significant variables.

It is clear from the study that that male HHHs gain higher profit (30 %), realize less profit loss (11.7 %), and remain in significantly higher PE (29.2 %) level than those households headed by women. A similar result was observed in households with concrete roof, access to irrigation facility, and access to a tractor (Table 6.4). The majority of households have houses with grass or mud-tile as roofing materials, and this category of households operate in 4.8 % less PE than those having a better-quality house, such as those having a concrete roof. The influence of irrigation facility is even higher (15.27 %) than that of house type and tractor access (10.5 %).

Moreover, this study shows the significantly positive role of incentive variables on PE. The variables used to estimate incentive index (Table 6.2) are further summarized with respect to observed profit, profit loss, and PE (Table 6.5). Households getting a higher output price, training, credit facility, source seed, and profit in seed processing are at higher PE levels as compared to their counterparts. However, the influence of credit (19.7 %), source seed (29.9 %), and profit at seed processing (17.7 %) remain rather higher as compared to the others. In spite of the aforementioned economic benefits, households realize social benefits (prestige) from seed production, and those realizing a social benefit have higher profit and operate at a higher profit efficiency level. Households declare that becoming members of SPOs would help farmers to strengthen their relationship with other service providers, which might have implications for accessing agricultural inputs and training (Khanal and Maharjan 2014).

In addition to the incentive indicators already discussed, seed-producing households realize the sense of empowerment by engaging in different types of training, visits, demonstrations, interactions, etc., and the outcomes of the empowerment could be translated into PE. As given in Table 6.6, the majority of households are aware of source seed, meaning that they know growing source seed is required for producing improved or truthfully labeled seed. These households fetch 21.4 % higher PE than those not aware of source seed. Similarly, it was found that households involved in rogueing (removal of unwanted plants from fields) by themselves, and those receiving rogueing supports from technical subcommittee members of their organizations, realize 12.5 % and 12.8 % higher PE, respectively, than their counterparts. In addition, seed laboratories owned by government

[2] Profit loss $=$ Maximum possible profit $-$ observed profit, and maximum possible profit $=$ Observed profit$/$PE

Table 6.4 Comparison of profit, profit loss, and profit efficiency in structural variables

Variables	*n*	Observed profit (NRs ha^{-1})	Profit loss (NRs ha^{-1})	Profit efficiency (%)
House types				
Concrete roof	54	48,584	56,381	44.1
Others	94	31,738	51,651	39.3
p value		0.021**	0.214	0.097*
Irrigation facility				
Yes	92	53,492	52,517	46.87
No	56	17,837	54,790	31.6
p value		0.000***	0.127	0.001***
Tractor use				
Yes	130	42,186	55,258	42.3
No	18	11,710	39,792	31.8
p value		0.001***	0.06*	0.09*

*, **, *** indicate difference between categories by 10 %, 5 %, and 1 %, respectively

Table 6.5 Comparison of profit, profit loss, and profit efficiency of incentive variables

Variables	*n*	Observed profit (NRs ha^{-1})	Profit loss[4] (NRs ha^{-1})	Profit efficiency (%)
Price				
Price	24	52,779	47,420	46.4
No	124	34,974	54,530	40.0
p value		0.0773*	0.334	26.3
Training				
Training	120	41,419	53,428	43.0
No	28	23,089	53,156	33.0
p value		0.0299**	0.324	0.058*
Loan				
Yes	72	47,814	58,609	43.1
No	76	29,338	48,420	39.1
p value		0.0082***	0.059*	21.4
Source seed				
Yes	143	40,451	52,564	42.3
No	5	1,837	76,636	12.4
p value		0.0014***	0.054**	0.001***
Profit in seed processing				
Yes	106	45,182	45,032	46.1
No	42	23,698	74,438	28.4
p value		0.0031***	0.000***	0.000***
Prestige				
Yes	125	40,280	53,676	41.4
No	23	26,862	51,753	39.1
p value		0.058*	0.324	12.4

*, **, *** indicate difference between categories by 10 %, 5 %, and 1 %, respectively

Table 6.6 Comparison of profit, profit loss, and profit efficiency of technical variables

Variables	*n*	Observed profit (NRs ha^{-1})	Profit loss (NRs ha^{-1})	Profit efficiency (%)
Knowledge of source seed				
Yes	125	43,701	51,703	44.7
No	23	16,246	61,613	23.3
p value		0.0008	0.170	0.0001***
Household engagement in rogueing				
Yes	136	41,319	52,827	42.1
No	12	12,279	59,612	29.6
p value		0.0043	0.256	0.049**
SPO facilitation in rogueing				
Yes	123	40,984	49,722	43.2
No	25	25,456	71,360	30.4
p value		0.031	0.0025	0.0203**
Field inspection by seed laboratory				
Yes	74	50,488	47,895	47.5
No	74	27,698	58,859	34.6
p value		0.0011	0.0423	0.0017***

*,**,*** indicate difference between categories by 10 %, 5 %, and 1 %, respectively

agencies provide technical facilitation in CSP, especially in disease and pest management as well as rogueing. As in the previous case, those receive technical facilitation from seed laboratories realize 12.9 % higher PE than those do not receive that support.

We had used cropping system as a dummy to represent diversification in R-W system, and the result shows that households adopting diversification realize a 0.8 % higher profit efficiency as compared to those not adopting this practice. As explained earlier, mungbean is the major crop adopted by farmers in the diversification scheme. So, higher profit efficiency of seed producers under crop-diversified schemes could be the result of nutrient supplementation from legumes in soil. Farmers grow mungbean utilizing residual moisture and nutrient, and the biomass is incorporated into the soil after one or two pickings of pods (about 80–90 days after seed sowing).

6.3 Discussion

Measuring profit efficiency is one of the popular approaches to understand farmers' performance in utilizing resources, especially among smallholder farmers in developing countries (Ali and Flinn 1989; Rahman 2003; Assa et al. 2012). This study has measured PE of farmers in rice and wheat seed production in a system perspective, which adds value in the existing efficiency literature because studies grasping the system perspectives are more likely to capture ground realities. This study shows the better performance of male-headed households in PE, which might be the result of better access of agricultural inputs such as chemical fertilizers to male-headed households. As learned from group discussions, availability of chemical fertilizer at appropriate times, volumes, and locations is a challenging issue in the study area. There is uncertainty about the availability of chemical fertilizers from the government mechanisms, and to address this issue, farmers go to India to buy chemical fertilizers. Because importing fertilizers from the Indian market without paying tax is not legally permitted, farmers, especially male-headed families, are in a better-off situation to bring fertilizer because of their better connection and risk-bearing capacity; this is further verified by the significantly higher amount of chemical fertilizers ($p = 0.041$) applied by male-headed households than those with a female head in a year, and the impact of chemical fertilizer on PE is positive (Table 6.3).

This study has recognized the importance of household infrastructural index in PE, and it indicates that household PE could be improved by the provision of physical assets including community seed storage, irrigation schemes, and tillage machinery. The relationship between community seed storage and PE can be explained from two dimensions. First, it improves seed quality by addressing the constraint of poor households who cannot sell seed in the market because of seed quality deterioration at storage (as is common in thatched-roof houses from seepage of moisture). Second, a seed storage facility might improve households' commercial orientation, which contributes to PE (Piya et al. 2012). Similarly, this study shows the positive impact of incentive variables: access to source seed, credit facility, and benefit from seed processing. Better performance of source seed accessed in PE might result from the linkage of source seed in increasing quantity and quality of seed production. Moreover, household realization of access to source seed as incentive while producing seed in an organized way implies that there is a challenge to obtain source seed, especially for those seed producers who are not engaged in groups/cooperatives, because government organizations (source seed suppliers) prioritize SPOs for source seed allocation assuming that source seed multiplied by SPOs could be collected back and utilized to address seed demand at the national level. The system of collecting seed back by project-implementing agencies is common in Asia and Africa (FAO 1998).

The positive role of credit facility on profit efficiency implies the necessity to integrate saving and credit schemes in the SPO development modality to enhance household PE, especially in those places where seed producers are in an early developmental phase. Previous studies on CSP have also highlighted the importance of credit in household performance in seed production and marketing (David 2004; Khanal and Maharjan 2010).

This study has also established that seed producer incentive can also be measured from the perspective of technical empowerment, and this benefit is ignored in the economic literature. However, it matters for the resource-poor farmers because farmers engaged in CSP are usually the members of farmers' groups and cooperatives, which were established not necessarily to maximize their profit from seed production but to enhance members' socioeconomic empowerment. The key indicators of empowerment measured in this study are knowledge about source seed, rogueing, and accessing services from extension agencies, and these variables have direct linkage with PE. The impact of these empowerment variables might not be limited to seed production activity only but could positively influence other on-farm and off-farm activities carried out by households.

6.4 Conclusion

This chapter measured the profit efficiency of seed producers involved in rice and wheat seed production using the stochastic frontier profit function. Variables considered in the efficiency measurement are labor, chemical fertilizer, livestock, and land. Some inefficiency-causing variables such as gender, education, dependency ratio, infrastructural index, incentive index, technical empowerment index, soil health index, and crop diversification index were also simultaneously modeled with major efficiency-determining variables. Results show that households are 41.1 % efficient in utilizing the aforementioned resources, and there is wide variability in the efficiency level among the households. This variation results from inefficiency generated by gender, infrastructure, incentive, and crop diversification indices. This study indicates the necessity to empower women farmers to enhance their efficiency in seed production and marketing. It also shows that PE could be improved through the provision of infrastructural facilities such as irrigation facility, seed storage, and agricultural machinery. Similarly, this chapter has validated that households realizing economic and social incentive by having engaged in SPOs are more efficient than those who did not realize these incentives. Moreover, there is a positive link between technical empowerment and PE, implying that incentive index (especially social benefit) and empowerment index could serve as important variables to measure the economic performance of farmers in CSP. The use of these indices would be much more useful where the seed CSP is in the early phase of its development.

References

Ali M, Flinn J (1989) Profit efficiency among Basmati rice producers in Pakistan Punjab. Am J Agric Econ 71(2):303–310

Assa MM, Edriss AK, Matchaya GC (2012) Unexploited profit among smallholder farmers in central Malawi: where are the sources? Int J Appl Econ 9(2):83–95

Battese GE, Coelli TJ (1995) A model for technical inefficiency effects in a stochastic frontier production function for panel data. Empir Econ 20:325–332

David S (2004) Farmer seed enterprises: a sustainable approach to seed delivery? J Agric Hum Values 21:387–397

FAO (1998) Developing seed security strategies and programs for food security in developing countries. In: Proceedings of the international workshop on seed security for food security: contribution for the development of seed security strategies in disaster-prone regions. Florence, 30 Nov–1 Dec 1997

Khanal NP, Maharjan KL (2010) Sustainability of community-based seed production enterprises in Nepal: institutional issues. Nepal Agric Res J 10:33–40. http://pustakalaya.org/eserv.php?pid=Pustakalaya:3772&dsID=NARC2010_NepalAgricultureResearchJournalVol10.pdf

Khanal NP, Maharjan KL (2013) Technical efficiency of rice seed growers in the Tarai region of Nepal. J Rural Prob 49(1):27–31

Khanal NP, Maharjan KL (2014) Adoption of improved rice varieties in the Tarai region of Nepal. In: Maharjan KL (ed) Communities and livelihood strategies in developing countries. Springer, Tokyo, pp 49–64. doi:10.1007/798-4-431-54774-7-4

Khanal NP, Khanal NN, Gurung GB, Thapa S, Gupta KP, Sherpa LT, Joshi KD, Harris D, Kumar Rao JVDK, Darai R (2008) Mungbean (*Vigna radiata* (L.) Wilczek) in cereal fallows: experience of farmers' participatory research and development activities in foothills and terai of Nepal. In: Kharkwal MC (ed) Proceedings of 4th international food legumes research conference, vol 2. Indian Society of Genetics and Plant Breeding, New Delhi, pp 67–76

Piya S, Kimanami A, Yagi H (2012) Comparing the technical efficiency of rice farmers in urban and rural areas: a case study from Nepal. Trends Agric Econ 5(2):48–60

Poudel DP, Choudhary K, Chowin R, Ghimire H (2003) Seed production and marketing through farmers' groups: case studies, Center for arid zone studies. University of Wales, Cardiff

Rahman S (2003) Profit efficiency among Bangladeshi rice farmers. Food Policy 28:487–503. Retrieved February 25, 2012, from http://www.sciencedirect.com/science/article/pii/S0306919203000745

Srinivas T, Bishaw Z, Rizvi J, Niane AA, Manan AR, Amegbeto K (2010) ICARDA's approach in seed delivery: technical performance and sustainability of village-based seed enterprises in Afghanistan. J New Seeds 11(2):138–163

Wang J, Cramer GL, Wailes EJ (1996) Production efficiency of Chinese agriculture: evidence from rural household survey data. Agric Econ 15:17–28

Witcombe JR, Devkota KP, Joshi KD (2010) Linking community-based seed producers to markets for a sustainable seed supply system. J Exp Agric 46(4):425–437

Chapter 7
Households' Behavior in Selling Rice and Wheat Seed in the Market

Abstract In community seed production, farmers produce seed at the household level but sell it to markets through their organizations. To maximize benefit at marketing, farmers tend to sell the maximum proportion of the seed produced in a household, but this does not hold true in subsistence farming. This chapter analyzes the impacts of socioeconomic variables in household seed-selling behavior in rice and wheat. A field survey was carried out in three Tarai districts of Nepal including 180 households, and data analysis was done using the Heckman selection model as it captures selection bias. Results showed that 65 % of the surveyed households sell rice seed in the market and the sold seed comprises 64.4 % of the total rice seed produced by a household. Similarly, for wheat 44.4 % of households participate in the market and sell 92.5 % of the produced seed. This chapter also identifies the significant roles of family labor, irrigation facility, operational land, training, share collection, and seed price in household behavior in selling rice and wheat seed in the market.

Keywords Rice-wheat seed selling • Household behavior • Heckman selection • Nepal

7.1 Introduction

In the community seed production (CSP) system, farmers realize benefits at two stages. In the production stage, they maximize the seed yield by proper allocation of resources, whereas in the marketing stage, they tend to increase seed price. Benefits realized in the first stage have been already discussed in Chaps. 5 and 6. This chapter discusses benefit realized by farmers at marketing stage. The marketing stage is mainly handled by seed producer organizations (SPOs), and normally these organizations are formed and managed by seed producers following democratic principles (David 2004; Witcombe et al. 2010). The marketing stage starts from the collection of seed from farmers to the SPO store, and it also involves several other activities such as seed processing, market research, and distribution. It is hypothesized that seed producers sell the maximum portion of their produced seed to SPOs soon after the crop harvest. The underlying reason behind it is that this approach increases the economy of scale of seed marketed by SPOs. However, this

N.P. Khanal, K.L. Maharjan, *Community Seed Production Sustainability in Rice–Wheat Farming*, DOI 10.1007/978-4-431-55474-5_7

assumption might not hold true in subsistence farming because part of the produced seed would be used as grain for food for the household or bartered with neighbors and relatives for social and economic reasons (Almekinders and Louwaars 1999; David 2004). So, it is quite difficult to quantify the effect of socioeconomic reasons in household seed-selling behavior directly (Setimela et al. 2004; Bishaw and van Gastel 2008; FAO 2010; Srinivas et al. 2010). Therefore, this chapter adopts an indirect approach to solve this problem by analyzing the impact of socioeconomic variables on farmers' behavior in selling seed in the market.

7.2 Methodology

7.2.1 Study Area and Sampling Technique

This study was carried out in three Tarai districts (Siraha, Chitwan and Kailali) of Nepal in November 2011. In these districts, rice-wheat system is the popular cropping system. Four SPOs with at least 2 years of experience in rice and wheat seed production were purposively selected. Fifteen households from each of the selected SPOs were chosen for field survey, making the total of 180 sample households.

7.2.2 Empirical Method

The Heckman selection model (Heckman 1979) was used for data analysis, and this model is considered preferable to ordinary least squares (OLS) if all the households do not participate in the market as it captures the selection bias. This model consists of two equations: the first equation is called the selection equation that gives the impacts of socioeconomic variables on probability of a household selling seed in the market, and the second equation, the outcome equation, indicates the impact of these variables on seed sold. These two equations were simultaneously modeled using the maximum likelihood method as it is more efficient than the two-step procedure (Nawata and Nagase 1996; Nawata 2004). To separate these two equations, the price of seed was used as an identifier variable in the selection equation. The outcome and selection equations are presented in Eqs. 7.1 and 7.2, respectively.

$$y_i = x_i\beta + \mu_i \tag{7.1}$$

$$z_i^* = w_i\alpha + \in_i \tag{7.2}$$

where, y_i is volume of seed sold in the market, $z_i{}^*$ is a latent variable. Similarly, x_i and w_i are the explanatory variables associated with outcome and selection equations, respectively. β is the vectors of coefficient, and μ_i and ϵ_i are the error terms.

The operational models of the outcome and selection equations are given in Eqs. 7.3 and 7.4, respectively.

$$\textbf{Ln seed sold} = \beta_0 + \beta_1\text{age of HHH} + \beta_2\text{education of HHH} + \beta_3\text{family labor} + \beta_4\text{ln off-farm income} + \beta_5\text{operational land} + \beta_6\text{livestock} + \beta_7\text{irrigation} + \beta_8\text{roof type} + \beta_9\text{share} + \beta_{910}\text{training} + \mu_i \quad (7.3)$$

$$\textbf{Market participation} = \alpha_0 + \alpha_1\text{age of HHH} + \alpha_2\text{education of HHH} + \alpha_3\text{family labor} + \alpha_4\text{ ln operational land} + \alpha_5\text{operational land} + \alpha_6\text{livestock} + \alpha_7\text{irrigation} + \alpha_8\text{roof type} + \alpha_9\text{share} + \alpha_{10}\text{training} + \alpha_{11}\text{seed price} + \in_i \quad (7.4)$$

where ln is log. Seed sold indicates the quantity of seed sold by households in the market, and it serves as a dependent variable in the outcome equation. It is possible that farmers sell seed not only to SPOs but also to others such as local farmers, agrovets (seed traders supplying agricultural inputs), development projects, and so on. But farmers, in group discussions, opined that they rarely sell seed directly to other users. Rather they sell seed to their SPOs (taken membership), and the SPOs handle the marketing process. So, SPOs are the first-hand buyers of rice and wheat seed produced by households. Similarly, market participation is the dependent variable in the selection equation, which shows whether farmers sell seed to SPOs (i.e., dummy variable that takes the value 0 or 1).

A total of 11 socioeconomic variables were chosen as explanatory variables considering economic theory, findings from previous literature, and experience of farmers as the combination of these strategies would help to draw the relevant variables for the study (Eq. 7.2). These variables include demographic (age and education of the household head, HHH, and family labor), economic (operational land, irrigation facility, off-farm income, livestock, and roof type), and institutional (training, collection of share in SPOs). The justification for the selection of these variables is given below.

The impact of age and education of HHH was hypothesized positive because age represents experience and education indicates the analytical capability, both of which might have positive impact on households' market participation and the volume of seed sold. Similarly, seed production of rice and wheat is carried out in rural areas where the majority of the work is done by the family members. Also, seed farming of these crops is seasonal in nature when most of laborers are busy in their own household activities. Even those wanting to hire laborers might not get them on time and could not operate field activities properly, which might influence the quantity and quality of seed. So, it was hypothesized that family labor (LFU)[1] would have positive impact on both market participation and volume of seed sold.

---

[1] LFU is a measure for labor force, where people from 15–59 years old regardless of their sex were categorized as 1 person = 1 LFU, but in case of children (10–14 years old) and elderly people (>59 years old) 1 person = 0.5 LFU

Operational land, irrigation facility (proportion of the total operational land with irrigation facility). and organic manure might have positive linkage on crop yield (Azam et al. 2012). Thus, these variables were assumed to have positive impact on household seed-selling behavior. Livestock (LSU)[2] was used as a proxy variable to represent the amount of animal manure applied in the field. Similarly, those having higher off-farm income might be less affected by cash/food shortage, especially from crop harvest until seed sold to SPOs, would be more motivated toward seed selling. Moreover, the SPOs are poor in physical structure (e.g., storage house, grading machine) in countries where seed industry is poorly developed, so they have to store seed at their personal houses for a few months after rice harvest until the SPO makes arrangement to store it in the common place/store (Fig. 7.1). Those having concrete-roofed houses would be more likely to be motivated toward seed selling as they could store the seed, maintaining its quality, for a longer time period than their counterparts. It means that in thatched-roofed households there might be a greater possibility of seed quality deterioration from leakage of moisture from

Fig. 7.1 Rice and wheat seed storage structure (From authors' photo bank)

[2] LSU is the aggregate of different types of livestock kept at household in standard units calculated using the following equivalents; 1 adult buffalo = 1 LSU, 1 immature buffalo = 0.5 LSU, 1 cow = 0.8 LSU, 1 calf = 0.4 LSU, 1 pig = 0.3 LSU, 1 sheep or goat = 0.2 LSU and 1 poultry or pigeon =0.1 LSU (Khanal and Maharjan 2014)

outside. Lower-quality seed might be rejected in the market, and even if accepted, households could get a lesser seed price.

Training and household's share (cash deposited at SPOs by farmers) are the two institutional variables considered in the study. It was assumed that those receiving training in any aspect of seed management (production, quality control, and marketing) might be better off both in market participation and seed-selling volume as training tends to enhance household motivation toward seed selling. Similarly, those who deposited cash at SPOs as share were assumed to have better performance in seed selling, because profit generated from marketing of seed could be distributed to farmers based on their proportion of the deposited share. The detail of dependent and explanatory variables used in the study is presented in Table 7.1.

Before running the Heckman selection model, data were validated for multicollinearity and heteroscedasticity. The Variance Inflation Factor (VIF) method was used to detect multicollinearity because this method is preferred over the correlation coefficient method (Pindyck and Rubinfield 1981). We did not find a problem of multicollinearity in the explanatory variables used in the model as the values are less than 10. The test for homogeneity of variance was conducted using the

Table 7.1 Description of explanatory variables and their expected sign

Variables	Definition	Mean ±SD	Expected sign
Age head of household (HHH)	Age of HHH in years	46.83±11.43	+
Education HHH	Formal schooling years of HHH	7.96±4.02	+
Family labor	Labor force unit (LFU) at household	3.44±1.44	+
Operational land	Total land for rice seed production (ha)	Rice: 0.95±0.36	+
		Wheat: 0.628±0.354	
Off-farm income	Annual household cash income from off-farm sources (NRs)	42,998±38,234	+
Irrigation	Operational land area under irrigation facility (%)	54.5±26.8	+
Livestock	Livestock standard unit (LSU)	3.86±5.77	+
Training	1= if household received seed management training, 0 = otherwise	0.783±0.413	+
Share	1= If farmers put share in the organization, 0 = otherwise	0.644±0.480	+
Roof type	1 = if households have concrete roof and 0 = otherwise	0.338±0.645	+
Seed price	Price of rice seed (NRs kg^{-1})	Rice: 18.02±2.81	+
		Wheat: 20.25±3.54	

Source: Survey 2011
1 US$ = NRs 82.96

Breusch-Pagan/Cook-Weisberg test for heteroscedasticity, and the null hypothesis of constant variances of the residuals was not rejected ($p > 0.25$) across the explanatory variables.

7.3 Results and Discussion

7.3.1 Seed Production and Selling

Farmers carry out rice seed production activity in most of their total operational land (79 %). However, there is variation in the proportion of land allocated for rice and wheat seed production. In rice, farmers grow seed on 0.95 ha, varying from 0.33 to 5 ha. Similarly, land allocated for wheat seed production averages 0.628 ha, ranging from 0.33 to 2.6 ha. Moreover, all the surveyed households were found to engage in rice seed production but 97 % of them were involved in wheat seed production. Rice yield (3,838 kg ha^{-1}) is quite higher than wheat yield (2,750 kg ha^{-1}) in the study area, and there is high variability in yield within each crop across the households. On average, households sell 2,087 kg rice seed and 1,684 kg wheat seed, and these figures represent 64.4 and 92.5 % of the total seed produced at household level in a year, respectively. Moreover, 65 % of the surveyed households sell rice seed in market but only 44.4 % of households sell wheat seed. It was found that households mostly sell seed through SPOs, and in some cases they sell seed as grain once seed quality has is deteriorated by intense rainfall during the crop harvest.

7.3.2 Summary of Explanatory Variables

Average age of HHH is 46.83 years but it varies from 17 to 75 years. Household heads undergo formal schooling for about 8 years, but 65 % of households have limited education, within the primary education level. Average off-farm income of households is NRs 42,998, which accounts for 31 % of the total annual cash income of the household. Public water canals connected with rivers are the major irrigation source, and households irrigate 54.5 % of their land using this source of irrigation. Similarly, the average LSU is 3.86, and major animals raised by farmers include cows, buffaloes, goat, and poultry (Table 7.1).

The majority of HHHs in the study area received seed management training (78.3 %) from government organizations and nongovernment organizations (NGOs). About one third of the households (33.8 %) had concrete-roofed houses. About two thirds of the households (64.4 %) have adopted the share collection practice in their organizations. The average seed price for rice and wheat is NRs 18 kg^{-1} and NRs 20 kg^{-1}, respectively. This price is the value of seed paid by SPOs to their members just after crop harvest. This rate is set by SPOs considering the grain price of the same commodity, the previous year's seed price, and the seed price set by a national seed company.

It was found that SPOs pay an additional NRs 2–3 kg^{-1} for seed as compared to grain, considering that the additional amount covers farmers' additional costs incurred for seed production as compared to the grain production of the same commodity.

7.3.3 *Output from Heckman Selection Model*

Table 7.2 presents the results from the Heckman selection model. The table shows that the variables chosen for the study fit this model well, as shown by the significant log-likelihood function ($p < 0.004$); that is, the coefficients of explanatory variables used in the models are significantly different from zero. Also, the log-likelihood ratio test rejected the hypothesis of the absence of correlation between the error terms of outcome and selection equations (rice: $p = 0.690$, $p = 0.027$; wheat: $p = 0.032$). This finding justifies the estimation of these two equations simultaneously using the Heckman selection model instead of OLS, which nullifies the censored observations. Because the equations were modeled using the maximum likelihood method, the coefficients of explanatory variables do not represent their average impact on the dependent variable. So, marginal impact of explanatory variables on dependent variables was estimated, and these impact values are used to discuss the degree of influence of these variables on the dependent variable. The study shows that the impact of most of the explanatory variables is in line with their hypothesized direction. However, the impact of some variables differs between the outcome and selection equations.

In rice, age of the HHH shows a significant positive impact on the volume of seed sold in the market. Households with a HHH 1 year older tend to sell a 1.9 % higher amount of seed than those with an average-aged HHH. The impact of this variable on market participation is not significant. There is significant positive impact of operational land on seed sold volume in the market, but its effect on market participation is not significant as in the age of HHH. An increase of 1 ha in operational land increases the seed sold volume by 6 %. Irrigation also showed significant positive impact on seed sold volume, which would be increased by 4.2 % with increase in the irrigated land by 1 %.

In contrast to the foregoing findings, livestock, training, and household share in SPOs showed significant positive impact on household participation in the market instead of volume of seed sold. As shown in Table 7.2, one unit increase in LSU leads to increasing the probability of a household selling seed in the market by 7.1 %. Similarly, there is significant positive impact of training on market participation. The households that attended training had increased probability to sell seed in the market that was is 15.5 % higher than the non-attendees.

Table 7.2 Impact of explanatory variables on outcome and selection equations

	Rice		Wheat	
Variables	Outcome equation	Selection equation	Outcome equation	Selection equation
Age HHH	0.020 (0.019)**	−0.002 (0.001)	−0.0207(0.0107)	0.010 (0.009)
Education HHH	0.027 (0.031)	0.021 (0.006)	0.034 (0.024)	0.036 (0.024)
Family labor	0.0213 (0.031)	0.027 (0.009)	0.119 (0.099)***	0.004 (0.003)
Operational land	0.07 (0.06)*	0.005 (0.0017)	0.474 (0.214)***	0.054 (0.025)
Off-farm income	0.4(0.2)	0.1(0.856)	0.079 (0.025)	0.308 (0.204)**
Irrigation	0.0765(0.042)*	0.112 (0.036)	0.876 (0.125)	0.793 (0.645)
Livestock	0.004 (0.002)	0.09 (0.071)*	0.004(0.003)	0.095(0.084)
Training	0.074 (0.0212)	0.182 (0.155)***	0.031(0.030)	0.044(0.032)
Share	0.081 (0.114)	0.220 (0.190)**	0.633(0.532)**	0.113(0.098)*
Roof type	0.271 (0.259)	0.033 (0.010)	−0.117(0.097)	−0.390(0.265)
Seed price	(0.11)*	0.08(0.071)*	(0.13)*	0.005(0.005)
Constant	6.433 (0.001)***	2.95(0.008)***	6.15	−4.93
Model characteristics	Wald test (χ^2, 10 = 17.66, log-likelihood statistics = 253.335, likelihood ratio test for $p = 0$ is 0.690**, Σ (sigma) = 1.098; λ (lambda) = 0.757; $n = 180$, censored observations = 63, uncensored observations = 117		Wald test (χ^2, 10 = 36.33***, log-likelihood statistics = −130.6, likelihood ratio test for $p = 0$ is 0.99**, Σ (sigma) = 1.058; λ (lambda) = 0.557; $n = 180$, censored observations = 63, uncensored observations = 117	

*, **, *** indicate significance at 10 %, 5 %, 1 % levels, respectively; figures in the parentheses are marginal values

The better performance of trained households in market participation might be the result of their better skills in seed quality management and commercial orientation (Witcombe et al. 2010). Similarly, there is significant impact of the household practice of share collection on market participation, and its impact on seed sold volume is also positive but not significant.

In wheat, family labor, operational land, off-farm income, share, and seed price show significant impact on household behavior in seed selling in the market. As in rice, the direction and magnitude of impacts of these variables on selection and outcome equation vary. One unit increase in family labor increases the probability of wheat growing for seed production by 0.3 %, but the magnitude of its impact is quite higher on volume of seed sold (9.9 %). This phenomenon could be explained from linkage of family labor with seed yield and its quality. Households with more family labor could plant wheat seed in timely fashion, and as a result this crop will be less affected by terminal drought and give a higher yield. Second, households with more family labor could adopt seed quality assurance measures such as rogueing (removal of unwanted plants from fields) in time. As a result, a greater portion of their produce is likely to be accepted by the technical subcommittee of the SPO as quality seed. The significant impact of operational land on volume of wheat seed sold would be their higher commercial orientation and greater amount of produce than their counterparts. Moreover, this study shows a significant positive impact of off-farm income on selecting a wheat crop for selling seed in the market, but its impact on volume of seed sold is not significant.

7.3.4 *Discussion*

Output price is the common indicator used to capture farmers' realized benefit level in marketing, especially in cash crops where the value chain is long and clear, which might not be applicable in seeds of open-pollinated seeds of food crop varieties. In this case, for example, if seed price increases farmers would sell a smaller portion of seed than their original plan because the same amount of benefit would be generated from selling a lesser seed quantity when selling at a higher price. However, it increases seed stock that farmers utilize for their different livelihood needs such as consuming seed in the form of food if not treated with pesticides in case of disasters. Again, it is quite difficult to separate benefit realized by households at two different phases (production and marketing) because the same households participate in these phases. Moreover, seed producers do not have full control of seed price determination as various internal and external factors interact in this process (Janvry et al. 1991; Benefica et al. 2006). Some of the internal factors contributing to raise price are strategies in marketing (processing, value addition, and distribution). However, adopting these strategies is not simple for small farmers residing in a marginal area (rain-fed areas) (Lanteri and Quagliotti 1997; Omit et al. 2009; Azam et al. 2012). In this context, the adoption of the indirect approach in analysis of household behavior is quite relevant, and identifying factors that motivate farmers in selling seed would be useful in designing appropriate extension policy as incentives could gear up seed industry growth in the country (Morris et al. 1998).

This study shows notable variation in market participation and the volume of rice and wheat seed sold across the households. In the case of rice, about two thirds of the surveyed households have participated in rice seed marketing and they sell a similar proportion of the total rice seed produced at the household. But in the case of wheat, a lesser proportion of households participate in the market but they sell a higher proportion of produce in market, possibly because there is a larger and more consistent demand for wheat seed than that of rice. Previous studies have also noted poor participation of farmers in the market in developing countries, especially in open-pollinated seed (Almekinders et al. 1994; Lal et al. 2009). Almekinders et al. (1994), and Wiggins and Cromwell (1995) argue that the poor performance of households in market is the consequence of poverty, poor marketing skills, and poor physical structures.

This study recognizes the importance of training on household behavior on rice and seed selling. In rice, a household that attended training sold 20 % more seed than their counterparts, whereas in wheat, households attending training sell 60 % more seed than their counterparts. Thus, the impact of training is much higher in wheat than in rice. Farmers take training on crop husbandry, seed quality maintenance, and marketing from projects implemented by NGOs and government agencies. Although 78.3 % of the households attended seed management training, only 65 % of the market participants received marketing training, mostly organized by NGOs. Moreover, some farmer received agricultural training from government agencies (32 % of the sample households), but the content of this training was focused on technical aspects of seed production with little or no information about marketing and entrepreneurship aspects, possibly the influence of the government policy document called 'District Seed Self-Sufficiency Program' on government-implemented projects. This policy document, which was designed and implemented by the Department of Agriculture in 1997, is focused on seed production and the marketing issue is ignored (Lal et al. 2009; Witcombe et al. 2010).

Another motivating factor for farmers to participate in seed marketing is the system of share holdings in SPOs. The share-collected households sell 26 % and 21 % more seed than their counterparts in rice and wheat, respectively. The higher motivation of share-collected households in selling seed would be their higher future benefit (at the marketing phase). In spite of this, only 50 % of households reported that their SPOs have distributed profit generated from seed marketing to their members, but some households (26 %) have been benefited by taking loans at cheaper interest from their SPOs to manage their household activities. In addition, the revenue collected at the organization level (from share collection and profit from seed marketing) has been mobilized for the development of a seed-processing facility (storage building and grading machine) at the organization level. The practice of share collection by households is considered important from the social perspective as well because it enhances household ownership and accountability toward SPOs. For instance, those with a larger amount hold higher voting rights in the decision-making process. This norm is similar to that of private companies and not to the general cooperative principle where 'one member, one vote' is applied regardless of the distribution of shares among the members (Acharya 2009). However, 75 % of the households included in this study are organized either in informal groups or cooperatives.

Moreover, the significant impact of livestock on household participation in the market might be its contribution to soil fertility and thereby crop yield. There is a positive correlation between LSU and crop yield both in rice ($r = 0.6$) and in wheat ($r = 0.58$). Livestock also contribute to a household's cash income, but only 20 % of farmers have received cash income from this subsector. The impact of operational land and irrigation facility is also significant and positive in household seed-selling behavior, implying that farmers with greater seed production area and better irrigation facility are more likely to sell greater quantities of seed in the market; this raises the concern of how to improve smallholders' performance in seed marketing. However, strategies such as timely payment for seed or provision of credit and an insurance system could address the small farmers' concerns and concerns of those growing rice under rain-fed conditions (Kugbei 2007).

7.4 Conclusion

This chapter analyzes households' behavior in selling rice and wheat seed in the market from a microeconomic perspective using the Heckman selection model. Results show that although there is great enthusiasm of households for rice and wheat seed production, many households have not shown good performance in marketing for various socioeconomic reasons. These constraints are small land size, limited irrigation facility, limited credit, limited training, and labor. It is also clear from this study that factors contributing to crop yield also influence a household's seed-selling behavior. These findings represent one approach to increase the business volume of SPOs. Increasing business volume is necessary to cover management costs of SPOs. These costs, at the time of the field study, were partially covered by development projects and the executive members were voluntarily engaged in their organization work. However, these organizations should cover their management costs by themselves to emerge as local enterprises in the future. Another approach for enhancing business volume of SPOs would be increasing their membership. However, this approach might increase management complexity, especially in the benefit-sharing process, because of the poor record-keeping system in SPOs. The associated issues in covering management cost are also related to the diversification of products supplied by SPOs. Given the limited demand for seed in the local market, it is quite difficult for SPOs to cover management costs while selling the seeds of only one or two crops.

References

Acharya BM (2009) Is cooperative really a democratic organization? Lessons learnt from the selected cooperative acts of the world. Annual bulletin. Cooperative Development Board, Kathmandu

Almekinders C, Louwaars N (1999) Farmers' seed production: new approaches and practices. Available via www.acss.ws/Upload/XML/Research/99.pdf. Accessed 12 Apr 2011

Almekinders C, Louwaars NP, Bruijin GH (1994) Local seed system and their importance for an improved seed supply in developing countries. Euphytica 78:207–216

Azam MS, Ima KS, Gaiha R (2012) Agricultural supply response and small holder market participation. The case of Cambodia. Available via http://www.rieb.kobe-u.ac.jp/academic/ra/dp/English/DP2012-09.pdf. Accessed 10 Jan 2013

Benefica R, Tschirley D, Boughton D (2006) Interlinked transaction in cash cropping economies: the determinant of farmers' participation and performance in the Zambezi River Vally of Mozambique. Contributed paper presented at the 26th international association of agricultural economists conference, Gold Coast, Australia, August 12–18, 2006

Bishaw Z, van Gastel AJG (2008) ICARDA's seed delivery approach in less favorable areas through village-based seed enterprises: conceptual and organizational issues. J New Seeds 9 (1):68–88

David S (2004) Farmer seed enterprises: a sustainable approach to seed delivery? J Agric Hum Values 21:387–397

FAO (2010) Promoting the growth and development of small seed enterprises for food security crops. Retrieved March 10, 2011, from www.fao.org/docrep/013/i1839e/i1839e00.pdf

Heckman JJ (1979) Sample selection bias as a specification error. J Econ 47(1):53–161

Janvry A, Fafchamps M, Sadoulet E (1991) Peasant household behavior with missing market: some paradoxes explained. Econ J 101(409):1400–1417

Khanal NP, Maharjan K (2014) Adoption of improved rice varieties in the Tarai region of Nepal. In: Maharjan KL (ed) Communities and livelihood strategies in developing countries. Springer, Tokyo, pp 49–64. doi:10.1007/798-4-431-54774-7-4

Kugbei S (2007) Seed economics: commercial consideration for enterprise management in developing countries. International Center for Agricultural Research in the Dry Areas Scientific Publishers, Jodhpur

Lal KK, Thapa M, Guenat D (2009) Review of seed projects in Nepal. Swiss Agency for Development and Cooperation, Bern

Lanteri S, Quagliotti L (1997) Problems related to seed production in African region. Euphytica 96:173–183

Morris ML, Rusike J, Smale M (1998) Maize seed industries: a conceptual framework. In: Morris ML (ed) Maize seed industries in developing countries. Lynne Rienner, Boulder

Nawata K (2004) Estimation of the female labor supply model by Heckman's two-step estimator and maximum likelihood estimator. Math Comput Simul 64:385–392

Nawata K, Nagase N (1996) Estimation of sample selection method. Economet Rev 15:387–400

Omiti JM, Otieno DJ, Nyanamba TO, Mccullough E (2009) Factors influencing the intensity of market participation by smallholder farmers: a case study of rural and peri-urban areas of Kenya. Available via http://ageconsearch.umn.edu/bitstream/56958/2/0301Omiti%20-%20FINAL.pdf. Accessed 4 Dec 2012

Pindyck RS, Rubinfield D (1981) Econometric models and economic forecasts. McGraw-Hill, New York

Setimela PS, Monyo E, Banziger M (eds) (2004) Successful community-based seed production strategies. CIMMYT, Mexico City

Srinivas T, Bishaw Z, Rizvi J, Niane AA, Manan AR, Amegbeto K (2010) ICARDA's approach in seed delivery: technical performance and sustainability of village-based seed enterprises in Afghanistan. J New Seeds 11(2):138–163

Wiggins S, Cromwell E (1995) NGOs and seed provision to smallholders in developing countries. World Dev 23:413–422

Witcombe JR, Devkota KP, Joshi KD (2010) Linking community-based seed producers to markets for a sustainable seed supply system. J Exp Agric 46(4):425–437

Chapter 8
Adoption of Soil Conservation Practices in Rice–Wheat Farming

Abstract Soil conservation practices harmonize economic, social, and environmental benefits for the rice-wheat farming system. This chapter analyzes the influence of socioeconomic factors on the adoption of existing soil conservation practices in community seed production. Potential environmental and social benefits are linked with economic gain as these benefits have time and risk questions to be realized by smallholder farmers. Data for the study were collected from 180 seed growers across the three Tarai districts of Nepal. Major soil conservation practices used in the rice-wheat cropping system are animal manure, zero tillage, and green manure. A multivariate probit model was run to identify the influence of socioeconomic factors on the adoption of zero tillage, green manure, and other practices. The result shows that households having less family labor, more operational land, and a higher risk aversion characteristic are more likely to adopt the zero-tillage practice. Similarly, those having access to an irrigation facility are more likely to adopt green manure crops. Irrigation facility, training, and risk aversion characteristic have a significant positive impact on the other practices whereas application of chemical fertilizers has significant negative impact on them.

Keywords Sustainability • Multivariate probit model • Green manure • Zero tillage

8.1 Introduction

Sustainability of the rice-wheat farming system has been a concern in recent years as long-term fertility trials show a stagnant or declining yield trend of these crops (Ladha et al. 2000; Tripathi et al. 2006; Erestein and Laxmi 2008). Similarly, the yield trend of these crops is fluctuating and almost stagnant, especially after 2000 in developing countries (FAOSTAT 2014). Although the exact causes of this problem are not yet fully understood, decreasing organic matter, increasing costs of cultivation, and scarcity of irrigation water are notable contributory factors (Timsina and Connor 2001; Erestein and Laxmi 2008). Moreover, being an intensive cropping system, the rice-wheat farming system is considered to have significance in emitting greenhouse gases such as methane (CH_4) and nitrous oxide (N_2O) (Lal

N.P. Khanal, K.L. Maharjan, *Community Seed Production Sustainability in Rice–Wheat Farming*, DOI 10.1007/978-4-431-55474-5_8

et al. 1998; Khanal et al. 2012). The aforementioned challenges of rice-wheat farming system apply to the community seed production (CSP) system as well.

Soil conservation practices (SCPs) have been found to contribute to both of these objectives by contributing economic, environmental, and social benefits to farmers. Different types of benefits rendered by SCPs are given in the reviews by Erestein and Laxmi (2008), Ereistein (2009), and Khanal et al. (2012). Economic benefit is related to increase as well as stability of crop yield because soils with high organic matter have a more stable structure and can resist shocks such as intense rainfall (Fig. 8.1). Similarly, organic soil reduces the cost of cultivation and increases the rental value of the land. Social benefit is related to leisure time (e.g., zero tillage reduces the labor requirement), and environmental benefit is related to maintaining environmental quality (soil, air, water, and biodiversity). Despite these notable benefits of SCPs, farmers need to choose an appropriate bundle of SCPs across the diverse socioeconomic contexts to sustain crop yield. Most of the previous studies on SCPs are about zero tillage, focusing on maintaining rice and wheat crop residues on the soil surface (Belknap and Saupe 1988; Lal et al. 1998; Erestein and

Fig. 8.1 An experiment comparing the stability of natural (forest) soil (*b*) and cultivated soil (*a*). Procedure: (1) Use three test tubes, filled with water to three fourths of length; (2) collect soil from cultivated field (*C*) and from forest area (*F*); (3) grind soil and remove plant debris; (4) put 50 g soil from C and F into test tubes (**a** and **b,** respectively), stir for 1 min with rod or spoon; (5) compare color of soil solutions after 5 min. Result: Most of the soil in test tube *a* is dissolved making the solution opaque, but it is opposite in test tube *b*, where most of the soil is deposited at the test tube bottom. Test tube *c* is the control (water only)

Laxmi 2008), and using seed drills for wheat planting in the mulched fields. These studies might have less policy implication for the small/subsistence farmers where there is a tradition of using residues from rice and wheat as feed for animals, fuel for household energy, and as thatching material. Small holders rarely adopt the tractor-drawn zero-till seed drill method unless there is development of local entrepreneur ship for production and supply of repair and maintenance of the machinery (Erestein 2009). Rather, the zero-tillage (ZT) practice has been traditionally adopted in the form of surface seeding (broadcasting of seed manually) of wheat in rural areas of developing countries (Tripathi et al. 2006; Gupta and Sayer 2007).

Previous studies on soil conservation issues have identified ZT (Granatstein et al. 1987; Ladha et al. 2000; Shah et al. 2011), green manure (Dahal et al. 1993; Devkota et al. 2006; Pandey et al. 2008), the system of rice intensification (SRI) (Upreti 2008), botanical pesticides, animal manure, and compost (Lal et al. 1998) as appropriate SCPs for small farmers. However, these studies mainly focused on technical matters and are based on researcher-controlled experiments, neglecting the influence of socioeconomic factors for their adoption. This chapter aims to contribute to the existing knowledge gap by presenting the influence of these factors in the adoption of SCPs.

8.2 Conceptual Framework

As already discussed, SCPs provide different economic, social, and environmental benefits to farmers, but the consideration of economic benefit might be more appropriate for small farmers because environmental and social benefits to be realized by this category of farmers are prone to time and risk questions and are not easily visible (Lee 1980). However, measuring economic benefits realized by farmers from SCPs is neither simple nor direct. Farmers could realize different economic benefits from the same SCP. For example, zero tillage contributes in water saving, reducing cost of cultivation, and building up soil organic matter. However, it is quite difficult to analyze how farmers prioritize these benefits. According to microeconomic theory, farmers as profit maximizers tend to innovate technologies in the process of utilizing the existing SCPs continuously, which makes it more difficult to put the long-term benefits to be realized by farmers adopting SCPs into utility functions. Moreover, criteria set by researchers and policy makers about economic benefits to be realized by farmers from SCPs might be different from those of the farmers. Further, farmers adopt these practices considering rice and wheat as a system, meaning that if SCPs are applied to one crop, farmers think that succeeding crops could be benefited by the SCPs (Kassie et al. 2012) as a consequence of improved soil quality. In this context, adoption studies carried out from the perspective that the farmers adopt these practices by perceiving potential costs and benefits from available SCPs in the cropping system could be more logical (Tripathi et al. 2006; Kassie et al. 2012). According to the adoption and diffusion theory (Rogors 1995), perception is a step in the technology adoption process, and it affects other steps of technology adoption. Perception is influenced by various demographic (age, education, labor, attitude), biophysical (disease, pest, climate stress, field characteristics), and economic, social, and institutional (land tenure, linkage) factors (Shiekh et al. 2002; Erenstein and Laxmi 2008; Kassie et al. 2012).

8.3 Methodology

The study was carried out in three Tarai districts: Siraha, Chitwan, and Kailali, representing eastern, central, and far-western development regions. The Tarai region, also known as the food basket of Nepal, is the major domain of the rice-wheat farming system that accounts for 66.2 % of cultivated land and 71.4 % of the production volume of these crops. Four seed producer organizations (SPOs) having at least 2 years of experience in rice and wheat seed production, and registered at District Agriculture Development Offices (DADOs) of the concerned districts, were selected. Fifteen households in each of the organizations were randomly chosen for the household survey, making a total sample size of 180. Group discussions were also held at SPO and DADO levels to complement the data collected from the household survey.

8.4 Empirical Model

8.4.1 Theoretical Concept

This chapter uses the multivariate probit (MVP) model to analyze the impact of socioeconomic variables for the adoption of different SCPs. This model fits in the analysis because the dependent variable is multivariate, binary, and factors of the dependent variables are correlated. This model assumes that given a set of explanatory variables the multivariate response is an indicator of the event with some unobserved latent variables (Z), assumed to arise from the multivariate normal (Gaussian) distribution, and falls within a certain interval (Tabet 2007). As discussed in the conceptual framework, farmers could integrate different SCPs to address their economic goal considering the cropping system perspective, which justifies the modeling of SCPs simultaneously using the MVP model rather than the univariate probit model. Although the multinomial logit model has been used to analyze the similar data in the literature, the MVP is more suited for correlated binary dependent variables that are not mutually exclusive (Shiekh et al. 2002; Young et al. 2009). Also, the MVP model relaxes the independence of the irrelevant alternatives property assumed by the logit model (Tabet 2007). This model was also used by previous researchers (Cappelari and Jenkins 2003; Kassie et al. 2012) to analyze the impacts of socioeconomic factors in household decisions for the adoption of SCPs. Theoretically, the MVP model can be presented as given in Eqs. 8.1 and 8.2:

$$Y_{ij}\begin{pmatrix} 1 \text{ if } Z_{ij>0} \\ 0 \text{ if no} \end{pmatrix} \tag{8.1}$$

$$Z_{ij} = x_i\beta + \varepsilon_i \tag{8.2}$$

where Y_{ij} is the binary dependent variable taking value 0 or 1 on the ith households and jth options in the dependent variable. Similarly, Z is the vector of latent variable, and β is a matrix of the regression coefficient associated with explanatory variables (x). Moreover, ε is a vector of the residual error term distributed as multivariate normal distribution with zero mean and unitary variance, $\varepsilon_i \tilde{} N(0, \Sigma)$, where Σ is the variance–covariance matrix having value 1 on the leading diagonal. The off-diagonal element in the correlation matrix, $\rho_{kj} = \rho_{jk}$, represents the unobserved correlation between the stochastic elements of the jth and kth items. The relationship between Z_{ij} and ρ_{jk} is given by the likelihood of the observed data that can be obtained by integration over the latent variables Z (Eq. 8.3).

$$p(Y_{ij} = 1/x_i, \beta, \Sigma) = \int A_{ij}, \ldots, \int A_{i1}\phi_r(Z_{ij}/X_i, \beta, \Sigma)dZ_{ij} \tag{8.3}$$

where A_{ij} is the interval (0,∞) if $Y_{ij} = 1$, and the interval (−∞, 0) otherwise.

Similarly, $\phi_j(Z_{ij}/X_i, \beta, p_{ij})$ is the probability density function of the standard normal distribution. The study uses the simulated maximum likelihood (SML) method using the Geweke Hajavassiou-Keane (GHK) simulator in STATA developed by Cappelari and Jenkins (2003) to estimate the MVP model. According to Cappelari and Jenkins (2003), the SML simulator tends to be consistent once the number of observations and number of draws tend to be infinitive. The number of draws is considered to be the square root of the sample size if the latter is 1,000 and above. However, for a small sample size the number of draws should be increased from its default number (5) to enhance the precision of the coefficient. So, the number of draws was set as 100 while estimating the model.

As per the regression rule, diagnostic tests were carried out to check the heteroscedasticity and multicollinearity problems in the data. For this, the same set of socioeconomic variables was regressed against the choice dependent variables individually using the ordinary least squares (OLS) technique. The variation inflation factor (VIF) test was carried out to check multicollinearity among the variables and did not find a problem as the values for explanatory variables across the three equations was less than 10. Similarly, the Breusch-Pagan/Cook-Weisberg test showed that the selected data set was free from heteroscedasticity as the null hypothesis of constant variances of the residuals was not rejected in all the tested equations.

8.4.2 Variables and Operational Model

A total of five SCPs [compost, zero tillage (ZT), green manure, system of rice intensification (SRI), and botanical pesticides] were found to have adopted by the farmers in seed production. Among these practices, ZT was adopted only in wheat whereas all other practices were adopted targeting mainly for rice. The SRI consists of a package of practices, such as planting seedlings 10 to 14 days old, maintaining wider spacing, provision of major nutrients from organic matter, and drainage (Upreti 2008). However, none of the farmers was found to have adopted the complete package of practices of SRI. So, those who adopted single seedlings while rice transplanting with any one of the foregoing practices were considered adopters of SRI. The variable animal manure was dropped from the analysis as all the farmers were found to adopt this practice. Then, the remaining five variables were used as dependent variables in the model. Again, the number of farmers adopting botanical pesticides, SRI, and compost were limited (Table 8.1). Thus, these three variables were combined under the name 'others,' and finally this variable and two other variables (green manure and zero tillage) were used as dependent variables in the model. The operational model used in this chapter is given in Eq. 8.4.

$$\begin{aligned}&(\text{Zero tillage} = \beta_{a0} + \beta_{a1}\ \text{age} + \beta_{a2}\ \text{education} + \beta_{a3}\ \text{family labor} + \beta_{a4}\ \text{income} + \beta_{a5}\ \text{operational land}\\&+ \beta_{a6}\ \text{livestock} + \beta_{a7}\ \ln \text{fertilizer cost} + \beta_{a8}\ \text{irrigation} + \beta_{a9}\ \text{diversification index (DI)} + \beta_{a10}\ \text{training} + \varepsilon_a)\\&(\text{Green manure} = \beta_{b0} + \beta_{b1}\ \text{age} + \beta_{b2}\ \text{education} + \beta_{b3}\ \text{family labor} + \beta_{b4}\ \text{income} + \beta_{b5}\ \text{operational land} +\\&\beta_{b6}\ \text{livestock} + \beta_{b7}\ \ln \text{fertilizer cost} + \beta_{b8}\ \text{irrigation} + \beta_{b9}\ \text{DI} + \beta_{b10}\ \text{training} + \varepsilon_b)(\text{Others} = \beta_{c0} + \beta_{c1}\ \text{age} +\\&\beta_{c2}\ \text{education} + \beta_{c3}\ \text{family labor} + \beta_{c4}\ \text{income} + \beta_{c5}\ \text{operational land} + \beta_{c6}\ \text{livestock} + \beta_{c7}\ \ln \text{fertilizer cost} +\\&\beta_{c8}\ \text{irrigation} + \beta_{c9}\ \text{DI} + \beta_{c10}\ \text{training} + \varepsilon_c)\end{aligned} \tag{8.4}$$

Here, Ln = log, β_a, β_b, and β_c are the vectors of the coefficients of explanatory variables related to zero tillage, green manure, and other practices, respectively. The explanatory variables used in the study were chosen on the basis of economic theory and previous studies. However, to capture the field reality, variables were validated with farmers before implementing the field survey. The literature shows that demographic, economic, and institutional variables might influence a household's decision for adopting SCPs, but the hypothesized relationship of the variables with adoption choices is specific to the conservation issue and locality (Ereistein 2009; Kassie et al. 2012).

The explanatory variables are classified as demographic, economic, and institutional. The summary of the variables and their hypothesized relationship with the dependent variables is given in Table 8.2. The demographic variables included in the study are age and education of household head (HHH) and family labor. Age and education were hypothesized to have positive influence on the adoption of all dependent variables as they contribute to the human capital (Ervin and Ervin 1982), whereas the influence of family labor on ZT was assumed negative as this technique saves labor and allows farmers timely wheat sowing (Tripathi et al. 2006).

Table 8.1 Summary of soil conservation practices

Categories	Overall	Chitwan	Siraha	Kailali	*p* value
Zero tillage					
Adopters	84 (47)	17 (28)	30 (50)	37 (62)	0.001
Non-adopters	96 (53)	43 (72)	30 (50)	23 (38)	
Compost					
Adopters	36 (20)	6 (10)	3 (5)	27 (45)	0.000
Non-adopters	144 (80)	54 (90)	57 (95)	33 (55)	
System of rice intensification (SRI)					
Adopters	22 (12)	3 (5)	4 (7)	15 (25)	0.001
Non-adopters	158 (88)	57 (95)	56 (93)	45 (75)	
Botanical pesticide					
Adopters	50 (28)	20 (33)	5 (8)	25 (42)	0.000
Non-adopters	130 (72)	40 (67)	55 (92)	35 (58)	
Green manure					
Adopters	109 (51)	31 (52	32 (53)	46 (77)	0.007
Non-adopters	71 (39)	29 (48)	28 (47)	14 (23)	

Source: Field survey 2011
Figures in the parentheses indicate percentage

The economic variables included in the model are income, operational land, livestock, fertilizer cost, irrigation, and the variety diversification index (VDI). The income variable included in this analysis represents the proportion of income generated from farming, and it was hypothesized to have positive impact across all the SCPs. As shown in Table 8.2, operational land was hypothesized to have positive influence across the dependent variables as this category of households has higher motivation to maximize crop production utilizing different resources (Shiekh et al. 2002). Household access to an irrigation facility was considered negative in the case of ZT because farmers normally use ZT in rain-fed areas, utilizing the residual moisture retained in the soil from the monsoon season (Tripathi et al. 2006). However, the impact of irrigation was hypothesized positive on 'green manure' and 'other practice' as irrigation might influence households for the adoption of these practices. The influence of livestock was assumed neutral to ZT and other practice, but it was hypothesized negative in case of green manure, because it might not be necessary to apply animal manure when green manure is applied in the field, as green manures are rich in plant nutrients. Similarly, the influence of fertilizer cost was assumed neutral to ZT but negative in green manure and other practices. Moreover, an irrigation facility was hypothesized as positive for green manure and other practices but negative for ZT.

Diversification of crop varieties is the common measure to address risks among smallholder farmers. It was hypothesized that higher risk averters are more likely to adopt SCPs because SCPs also contribute in enhancing the diversity of soil microbes (Belknap and Saupe 1988). The VDI was calculated as a ratio of number of rice varieties grown by a household to the total number of rice varieties grown in

Table 8.2 Summary of explanatory variables and their expected sign

Variables	Definition	Mean± SD	Expected sign
Age of HHH	Age of household head (years)	46.83 ± 11.43	+ ve to ZT and −ve to others
Education of HHH	Formal schooling attended by HHH (years)	7.96 ± 4.02	+ve to all
Family labor	Labor force unit at household (LFU)[a]	3.41 ± 0.37	− ve to ZT and +ve to others
Income	Proportion of income received from farming	0.40 ± 0.25	+ve to all
Operational land	Total operational land of households (ha)	1.15 ± 0.90	+ve to all
Livestock	Livestock standard unit (LSU)[b]	3.86 ± 4.88	−ve to green manure and +ve/−ve to others
Fertilizer cost	Chemical fertilizers households (NRs/ha/year)	6,649.1 ± 4,850.4	−ve to all
Irrigation	1 = access to canal irrigation, 0 for otherwise	0.55 ± 0.49	Same as family labor
Diversification index	Variety diversification index in rice	0.19 ± 0.07	+ve to all
Training	1 = Attended agriculture training, and 0 for otherwise	0.783 ± 0.41	+ve to all

ZT zero tillage, *SD* standard deviation, 1 US$ = NRs 82.96

[a]LFU is the measurement of labor force, where people from 15 to 59 years old regardless of their sex were categorized as 1 person = 1 LFU, but in case of children (10–14 years old) and elderly people (>59 years old) 1 person = 0.5 LFU

[b]LSU is the aggregates of different types of livestock kept at household in standard unit calculated using the following equivalents; 1 adult buffalo = 1 LSU, I immature buffalo = 0.5 LSU, 1 cow = 0.8 LSU, 1 calf = 0.4 LSU, 1 pig = 0.3 LSU, 1 sheep or goat = 0.2 LSU and 1 poultry or pigeon =0.1 LSU (Khanal and Maharjan 2014)

the concerning SPOs. The institutional variable considered in this study is a household's linkage with government and nongovernment organizations. These organizations provide training to the seed growers in various dimensions of seed production including SCPs. So, households attending agricultural training were used as a proxy variable to represent their linkage with these organizations. It was assumed that trained households are more likely to adopt SCPs because of the knowledge and experience they acquire from training.

8.5 Results and Discussion

8.5.1 Descriptive Analysis

As shown in Table 8.1, 46.67 % of the households adopt ZT practice in wheat, and the proportion of households adopting this practice is higher in Kailali as compared to Chitwan or Siraha. Overall, 50.56 % of the households have adopted green manure, and the category of the green manure adopters is also higher in Kailali as compared to other districts. The green manure crops grown by farmers are maize (*Zea mays*), *Dhaincha* (*Sesbania* spp.), and *Til* (*Sesamum indicum*) in Chitwan, whereas it is mungbean (*Vigna mungo*) in Siraha and Kailali. Few households (20 %) were found to adopt compost, and Kailai district has the majority of the adopters.

Farmers prepare compost mainly using the biomass from forest (dried fallen leaves), and it is supplemented with an effective microorganism and/or manure slurry to accelerate the decomposition of leaves. Only 12 % of the households adopt SRI, and the proportion of SRI adopters was also higher in Kailali as compared to the other two districts. Similarly, only 27 % of the households use botanical pesticides to control pests/diseases in rice; this does not mean that the remaining households use chemical pesticides, as only 3 % of households were found to use chemical pesticides in rice for weed control.

As discussed previously, we have combined SRI, compost, and botanical pesticides under the group 'other practice' in this research. The summary statistics show that overall 44 % of households adopted 'other practices,' and the proportion of adopters is higher in Kailali (76 %) as compared to Chitwan (40 %) and Siraha (16 %).

Table 8.2 presents the summary of explanatory variables used in the analysis with respect to their mean and standard deviation. In case of dummy variables, the mean value indicates the percentage of households adopting the practices. For example, the mean value for irrigation facility is 0.55, meaning that 55 % of the households have access to a public irrigation source. Farmers primarily use chemical fertilizer in rice, and its use is quite lower than the national recommendation for rice (100:30:30 N:P:K kg ha^{-1}) (Tripathi et al. 2006). Overall, farmers apply major plant nutrients at 24.7 kg N, 20.9 kg P, and 17 kg K ha^{-1}. These nutrients are supplied through different chemical fertilizers such as urea (46 % N), diammonium phosphate (18 % N and 46 % P), and muriate of potash (60 % K).

8.5.2 Results from Multivariate Probit Model

The study shows that the direction of response of most of the socioeconomic variables is as per the expectation with few exceptions (Table 8.3). The significant likelihood function, as given by the Wald test, indicates that the variables chosen in the study fit the model well. Also, the likelihood ratio test rejected the hypothesis of the independence of the error term of individual equation ($p = 0.0001$). Households having more family labor are significantly less likely to adopt the ZT practice, which might be because of the ability of households having more family labor to plant wheat on time, as wheat planting after the end of November could decrease the crop yield (Aslam et al. 1993).

Table 8.3 Impact of socioeconomic variables on the adoption of soil conservation practices

Variables	Zero tillage (1)	Green manure (2)	Other practices (3)
Age of HHH	0.007 (0.411)	−0.010 (0.242)	0.0019 (0.830)
Education of HHH	−0.010 (0.695)	−0.010 (0.706)	−0.022 (0.423)
Family labor	−0.015 (0.0675)*	−0.014 (0.680)	−0.014 (0.679)
Income	0.001 (0.337)	0.004 (0.745)	0.002 (0.546)
Operational land	0.380 (0.008)***	0.170 (0.206)	0.025 (0.835)
Livestock	0.001 (0.850)	−0.008 (0.274)	0.003 (0.638)
Fertilizer cost	0.001 (0.204)	−0.001 (0.108)	−0.001 (0.030)**
Irrigation	−0.073 (0.735)	0.394 (0.033)**	0.540 (0.013)**
Diversification index	2.933 (0.055)**	2.21 (0.745)	3.548 (0.024)**
Training	0.005 (0.984)	0.161 (0.500)	0.477 (0.064)*
Constant	−1.525 (0.021)**	−0.169 (0.155)	−1.704 (0.011)**

$\rho_{21}=0.496$ (0.001)***, $\rho_{23}=0.321$ (0.005)***, $\rho_{31}=0.302$ (0.006), $n=180$; Wald test log likelihood $=-318$, chi square (30): 49.35, $p=0.0014$; log-likelihood ratio test; $\rho_{21}=\rho_{21}=\rho_{21}=0$, chi-square $=27.72$, $p=0.0001$; $n=180$; number of draws $=100$; figures in the parentheses indicate probability values; *,**, ***, significance at 10 %, 5 %, and 1 %, respectively

This study also shows the positive linkage between ZT and wheat yield as ZT adopters received 8 % higher yield (Table 8.4) than the control group (those who do not apply any SCPs), but it does not indicate that the increased wheat yield was only the result of early planting of wheat (escape of terminal drought; i.e., desiccation of floral parts before maturity from high temperature) because ZT also contributes to the increase in soil organic matter. Granatstein et al. (1987) found 0.2 % increment in organic matter after the adoption of ZT practice continuously for 10 years in wheat. Similarly, Shah et al. (2011) also found better performance of ZT wheat when integrated with mulch. Perhaps because of improved soil quality, ZT adopters achieved 3.1 % higher rice yield as compared to the control group. Similarly, the study shows that households with greater operational holdings are significantly more likely to adopt the ZT practice (Shiekh et al. 2002). Moreover, households with higher VDI are also significantly more likely to adopt this practice. VDIs here represent the attitude of farmers toward risks, and it shows that the risk aversion characteristic could positively motivate farmers to adopt ZT. Farmers argued that the most important risk they assume to have reduced from adopting ZT is escape of the crop from terminal drought because it allows them to sow wheat seed on time, that is, within November. Belknap and Saupe (1988) also noted that planting of wheat after first week of December significantly decreases the wheat yield.

Similarly, green manure adopters achieved 24.8 % higher yield in rice as compared to the control group. This practice also has a complementary role in enhancing the productivity of zero tillage and other practices (Table 8.4), which might be the result of the improvement of soil quality.

In addition, even if green manure were applied only in rice, green manure adopters achieved 13.4 % higher yield in wheat as well. It clearly shows the role

Table 8.4 Rice and wheat yield under different soil conservation practices

	Rice (kg ha^{-1})		Wheat (kg ha^{-1})	
Conservation practices	Mean± SE	Over control (%)	Mean± SE	Over control (%)
Zero tillage (ZT)	3,420 ± 378	3.1	2,685 ± 167	8
Green manure (GM)	4,140 ± 228	24.8	2,818 ± 152	13.4
Other practice (OP)	3,745 ± 237	12.9	2,607 ± 237	4.9
ZT*GM	3,702 ± 200	11.6	2,620 ± 132	5.4
ZT*OP	3,748 ± 208	13.02	2,602 ± 223	4.7
GM*OP	4,036 ± 196	21.7	3,045 ± 221	22.5
ZT*GM*OP	4,322 ± 147	30.3	3,148 ± 153	26.68
Control	3,316 ± 129	–	2,485 ± 102	–

Source: Field survey, 2011
SE standard error

of green manure in enhancing the productivity of rice-wheat farming system as a whole.

Previously, researchers have also found positive results using green manure on rice and wheat yield as well as other characteristics of soil such as microbial diversity and aeration (Dahal et al. 1993; Devkota et al. 2006; Pandey et al. 2008). Farmers grow different green manure crops in their land in the spring season (April to June) after harvesting wheat and before transplanting the main season rice (July to October), and incorporate their green biomass into soil during the final land preparation for rice. Of the various socioeconomic variables tested for using green manure, only irrigation shows a significant impact. Thus, households having access to an irrigation facility (canal irrigation from a river/stream) are more likely to grow green manure crops, which shows that irrigation is one of the constraints for the adoption of green manure crops. Although there is ample opportunity for the use of underground water through tube wells for irrigating green manure crops, only 5 % of the adopters were found to have their own tube well, and farmers argued that the lower adoption of tube-well irrigation was caused by high installation costs, although 55 % of the households have an irrigation facility and farmers in the group discussion stated that water level in the irrigation canal is reduced by 50 % to 75 % in the spring season. As a result, many farmers have to wait for rainfall to sow a green manure crop, and sometimes they fail to sow seed for these crops. In contrast to ZT, most farmers (90 %) are aware of the role of green manure in the improvement of soil quality. Although there is no significant influence of chemical fertilizer on green manure adoption, the negative coefficient of chemical fertilizer to some extent indicates that households adopting more chemical fertilizer are less likely to adopt a green manure crop.

As in ZT and green manure, other practice adopters received higher yield both in rice (12.9 %) and wheat (4.9 %) as compared to the control group (Table 8.4). The reason for higher crop yield in improved practice adopters might be the contribution of the practice to enhance soil organic matter. The adoption of an improved practice is influenced by irrigation, training, fertilizer cost, and variety diversification. Thus, households having an irrigation facility, having taken agricultural training, and having higher risk aversion characteristics are more likely to adopt an improved practice, but those using more chemical fertilizer are significantly less likely to adopt an improved practice. This factor also justifies farmers' behavior in adopting an improved practice, considering the economic perspective, although the practices also contribute in saving water (e.g., SRI), and reducing pollution, but enhancing soil quality.

Furthermore, other socioeconomic variables not already discussed did not show any significant influence on the adoption of SCPs. However, these variables may also be important in a household's decisions.

8.6 Conclusion

This chapter shows that seed producers adopt different SCPs in seed production under the rice-wheat farming system. These practices include ZT, SRI, and improved practices, and the proportion of households adopting these practices is 46.67 %, 50.56 %, and 44 %, respectively. The influence of households' socioeconomic variables on the adoption of zero tillage, green manure, and improved practice in the rice-wheat farming system was analyzed using the MVP model. Although SCPs provide economic, social, and environmental benefits, economic consideration was adopted in the analysis because social and environmental benefits realized by small farmers have time and risk questions. There is a positive linkage of SCPs with crop yield in both rice and wheat, which justifies the use of economic considerations in the analysis, although SCPs impart other benefits as well. The results of the MVP model indicate that influence of socioeconomic variables varies across SCPs. Households with less family labor, more operational land, and higher risk aversion characteristics are more likely to adopt zero tillage. Similarly, those having an irrigation facility are more likely to adopt a green manure crop, which shows the importance of promoting mungbean in the irrigation domain. However, it is important to promote a basket of SCPs in the rice-wheat system and to raise people's awareness about the economic and ecological benefits rendered by SCPs through training.

References

Aslam M, Majid A, Hashmi NI, Hobbs PR (1993) Improving wheat yield in the rice-wheat cropping system of the Punjab through zero tillage. Pak J Agric Res 14:8–11

Belknap J, Saupe WE (1988) Farm family resources and the adoption of no-plow tillage in southwestern Wisconsin. N Cent J Agric Econ 10:13–23

Cappelari L, Jenkins SP (2003) Multivariate probit regression using simulated maximum likelihood. Stata J 3(3):278–294

Dahal KR, Mishra NK, Rajbhandary BP, Chaudhary NK, Prasad ML (1993) Green manuring studies on rice. Research report. Institute of Agriculture and Animal Science, Rampur, pp 40–43

Devkota KP, Yadav DN, Chaudhary NK, Dangol DR, Basnet KB (2006) Influence of spring season crop residue on productivity of rice-wheat cropping system. J Inst Agric Anim Sci 27:53–58

Erenstei O, Laxmi V (2008) Zero tillage impacts in India's rice-wheat systems: a review. Soil Tillage Res 100(1):1–14. Retrieved October 14, 2012, from http://www.mendeley.com/research/zero-tillage-impacts-in-indias-ricewheat-systems-a-review/

Erestein O (2009) Zero tillage in rice-wheat system of Indo Gangetic Plain: review of impacts and sustainability implications. A discussion paper. Retrieved 13 May 2011, from http://www.ifpri.cgiar.org/sites/default/files/publications/ifpridp00916.pdf

Ervin CA, Ervin DE (1982) Factors affecting the use of soil conservation practices. Land Econ 58:79–90

FAOSTAT (2014) Crop information 2011. Retrieved 11 Feb 2013, from http://faostat.fao.org/site/567/DesktopDefault.aspx?PageID=567#ancor

Granatstein DM, Bezdicek DF, Cochran VL, Elliot, Hammel J (1987) Long term tillage and rotation effects on soil microbial biomass, carbon and nitrogen. Biol Fertil Soils 5:265–270

Gupta R, Sayre K (2007) Conservation agriculture in South Asia. J Agric Sci 145(3):207–214. Retrieved October 15, 2012, from http://www.ifpri.cgiar.org/sites/default/files/publications/ifpridp00916.pdf

Kassie M, Jaleta M, Shiferaw B, Mmbando F (2012) Plot and household level determinants of sustainable agricultural practices in rural Tanzania. Environ Dev. Retrieved 12 Oct 2012, from http://www.rff.org/RFF/documents/EfD-DP-12-02.pdf

Khanal NP, Maharjan K (2014) Adoption of improved rice varieties in the Tarai region of Nepal. In: Maharjan KL (ed) Communities and livelihood strategies in developing countries. Springer, Tokyo, pp 49–64. doi:10.1007/798-4-431-54774-7-4

Khanal NP, Maharjan KL, Dangol DR (2012) Soil conservation practices for the sustainability of rice-wheat system in Nepal: a review. J Int Dev Coop 18(4):11–20

Ladha JK, Fischer KS, Hossain M, Hobbs PR, Hardy B (eds) (2000) Improving the productivity and sustainability of rice-wheat systems of the Indo-Gangetic Plains: a synthesis of NARS-IRRI partnership research. A Discussion Paper Series No. 40. International Rice Research Institute, Manila

Lal R, Kimble J, Follett R, Cole C (1998) The potential of U.S. cropland to sequester carbon and mitigate the greenhouse effect. Ann Arbor Press, Chelsea

Lee LK (1980) The Impact of landownership factors on soil conservation. Am J Agric Econ 62 (5):1070–1076

Pandey KR, Shah SC, Becker M (2008) Management of native soil nitrogen for reducing nitrous oxide emission and higher rice production. J Agric Environ 9:1–9

Rogers EM (1995) The diffusion of innovations. The Free Press, New York

Shah P, Dahal KR, Shah SK, Dangol DR (2011) Effect of tillage, mulch and time of nitrogen application on yield of wheat (*Triticum aestivum* L.). Agric Dev J 8:9–19

Shiekh AD, Rahman T, Yates CM (2002) Logit models for identifying the factors that influence the uptake of new no-tillage technologies by farmers in the rice-wheat and the cotton-wheat farming systems of Pakistan's Punjab. Agric Syst 75:79–95

Tabet A (2007) Bayesian inference in the multivariate probit model: estimation of the correlation matrix. Master's thesis, The University of British Columbia, Vancouver

Timsina J, Connor DJ (2001) Productivity and management of rice-wheat cropping systems: issues and challenges. Field Crop Res 69:93–132

Tripathi J, Chiranjivi A, Lauren JG, Duxbury JM, Hobbs PR (2006) Assessment of farmer adoption of surface seeded wheat in the Nepal Tarai. Rice-wheat consortium paper series 19. Rice-Wheat Consortium, New Delhi. Retrieved 10 Jul 2011, from gradeducation.lifesciences.cornell.edu/faculty/individual800

Upreti R (2008) System of rice intensification in the context of Nepalese rice production. Retrieved 10 Mar 2012, from sri.ciifad.cornell.edu/countries/nepal/nepalrptupprety04.pdf.
Young GE, Valdez A, Kohn R (2009) Multivariate probit models for conditional claim-types. Insur Math Econ 44:214–228

Chapter 9
Risk Management in Community Seed Production Under Rice–Wheat Cropping System

Abstract In community seed production, seed producers face various physical and financial risks, the impacts of which are reflected in profit from seed production. This chapter analyzes the profit risk incidence, risk gap, and risk severity of seed producers involved in rice and wheat seed production, using cross-sectional data collected from 180 households. Households that realized profit at less than average amounts were considered to be at risk. The results show that 32.7 % of the surveyed households are at risk, and there is a 65.23 % risk gap across the households under this category. The risk severity of these households is 42.54 %. Share collection and crop diversification are the common risk management practices adopted by surveyed households. Household behavior in adopting these practices has also been analyzed using univariate probit model. It is clear from the analysis that households with a younger household head, and those associated with the Brahmin caste, are more likely to adopt the share collection practice. Similarly, households associated with a less-educated household head, and smaller operating landholders, are more likely to adopt crop diversification practices. Moreover, those that are better off in physical assets, that realized higher incentive, and with better technical capacity are more likely to adopt both these risk management practices.

Keywords Profit risk • Rice-wheat system • Univariate probit • Crop diversification • Share collection

9.1 Introduction

Farmers involved in community seed production (CSP) represent smallholders who are residing in marginal areas and engaged in seed production of open-pollinated varieties of food crops (Khanal and Maharjan 2010). These farmers confront risks from various factors such as finance, climate, pests, diseases, and market and human factors (Kugbei 2007; Srinivas et al. 2010). The financial risk is associated with uncertainty/problem in accessing credit to implement seed production activities on time, and this might have a significant negative impact on quantity and quality of seed production. Access to credit is also important after crop harvest because without a credit facility many smallholders might be compelled to sell seed as grain soon after the crop harvest (Khanal and Maharjan 2013). As a result,

N.P. Khanal, K.L. Maharjan, *Community Seed Production Sustainability in Rice–Wheat Farming*, DOI 10.1007/978-4-431-55474-5_9

farmers could not get the potential benefit they intend to get from seed marketing, that is, increasing output price by processing and value addition (Khanal and Maharjan 2013). Financial institutions consider it is risky to lend to smallholders, and these institutions are poorly developed in the rural areas of developing countries (Pradhan 2009).

Similarly, climate risk is associated with negative impact of climate variables such as rainfall and temperature on crop yield and output price. Generally seed producers intend to be benefited by extension agencies and policy makers. However, these farmers could face risks from these organizations because of their changing subsidy policies, accountability, and project implementation modalities (Almekinders et al. 1994). The presence of risks in CSP from these sources has given impetus to the development of a number of risk management strategies as these strategies could enhance resilience of CSP by improving farmers' adaptive and innovative capacity (McGuire and Sperling 2013). The common strategy for risk management for smallholder farmers is diversification of livelihood strategies including crop varieties. However, seed producers adopt different risk management strategies considering the possibility of risks appearing in both crop production and marketing phases; therefore, one strategy alone could not address the risks experienced by households in these two phases. This chapter analyzes profit risks in seed production under the rice-wheat farming system and farmers' behavior in adopting different risk management strategies.

9.2 Methodology

9.2.1 Study Area and Sampling Technique

This study was carried out in three Tarai districts of Nepal: Siraha, Chitwan, and Kailali. These districts capture the overall heterogeneity available in the Tarai region (plains area with altitudinal range from 70 to 650 m above sea level). Among the districts, Sirha lies in the eastern development region and is considered a drought-prone district. Located in the far-western development region, Kailali has been suffering from flood; whereas Chitwan lies in central Nepal and remains almost neutral in terms of drought and flood (MoE 2010). From each of these districts, four farmers' seed producer organizations (SPOs) with at least 2 years of experience in rice and wheat seed production were selected in consultation with district agriculture development offices. Then, 15 seed-producing households from each of the selected organizations were randomly selected to make the sample size of 180. From the selected households, data on household resources, and costs of production and profits from different crops grown under the rice-wheat farming system were measured.

9.2.2 Empirical Technique for Risk Estimation

We measured risk incidence, risk gap, and risk severity using profit variable (gross revenue less total variable costs involved in rice and wheat seed production). In agricultural production, risk can be measured by analyzing input price, output price, crop yield, and profit. However, the profit captures the essence of the other three variables. In this case, risk incidence is the proportion of households having gross profit less than the average gross profit of the total sampled households (Eq. 9.1). Similarly, risk gap is the extent to which individuals on average fall below the average profit line. In other words, risk gap is the average profit less actual profit for poor-performing seed producers whose profit remains below the average profit line. The risk gap can be expressed in terms of risk gap index (Eq. 9.2) as the proportion of risk gap in the population (those realizing profit above average have zero risk gap). It can also be taken as the cost of eliminating profit risk in CSP, because it shows how much to transfer from non-risk households to the risk ones. To address the issue of inequality among the households, risk severity index was also estimated (Eq. 9.3) by squaring the risk gap: this is simply the weightage sum of risk gap where the weights are the proportionate of risk gaps themselves.

$$\text{Risk incidence} = \frac{N_{\text{ba}}}{N} \tag{9.1}$$

$$\text{Risk gap} = \frac{1}{N}\sum_{i=1}^{N}\frac{G_i}{\dot{Y}} \tag{9.2}$$

$$\text{Risk severity} = \frac{1}{N}\sum_{i=1}^{N}\left\{\frac{G_i}{\dot{\text{Y}}}\right\}^2 \tag{9.3}$$

where N_{ba} is the number of households with gross profit below average, N indicates the sample size, G_i is the difference between observed profit and average profit $\left(\dot{Y} - Y\right)$, Y = observed profit of the surveyed household, and $\dot{\text{Y}}$ = mean profit of the surveyed households.

9.2.3 Empirical Technique for Analyzing Risk Management Strategies

General Background of the Model

The univariate probit (UVP) model was employed to analyze farmers' behavior in adopting risk management practices (RMPs) in seed production and marketing. These practices include share collection and crop diversification. It is believed that these practices could address risks coming from various sources. For example, the practice of collecting shares (practice of collecting cash at SPOs) is believed to address financial risk as those collecting shares in these organizations are more likely to access s credit facility from the respective SPOs. The share-collecting households could also minimize the human risk because of their empowerment as a result of a training facility. Similarly, the practice of crop diversification addresses the risks posed by climatic factors (Valandia et al. 2009). The crop diversification in this study indicates whether households grow other crops for seed production purposes in addition to rice and wheat. These RMPs are binary in nature, and farmers could adopt one or a combination of these strategies simultaneously to address risks coming from various sources. These two choices are not mutually exclusive, and this justifies the use of the probit model for data analysis (Cappelari and Jenkins 2003).

This model assumes that given a set of explanatory variables, the univariate response is an indicator of the event that some unobserved latent variables (Z), assumed to arise from the univariate normal distribution, fall within a certain interval (Tabet 2007). Also, the UVP model relaxes the independence of the irrelevant alternatives property assumed by the logit model (Shiekh et al. 2002; Young et al. 2009). Theoretically, the UVP model can be presented as given in Eqs. 9.4 and 9.5:

$$Y_i = \begin{pmatrix} 1 \text{ if } z_{i>0} \\ 0 \text{ if } z_{i\leq 0} \end{pmatrix} \tag{9.4}$$

$$Z_i = x_i\beta + \varepsilon_i \tag{9.5}$$

where Y_i is the binary dependent variable taking value 0 or 1 in the ith households. Similarly, z_i is the vector of the latent variable, β is a matrix of the regression coefficient associated with explanatory variables (X_i), and ε is a vector of the error. As per the regression rule, diagnostic tests were carried out to check the heteroscedasticity and multicollinearity problems in the data.

For this, the same set of socioeconomic variables was regressed against the dependent variable separately using the ordinary least squares (OLS) technique. A variation inflation factor test was carried out to check multicollinearity among the variables and did not find this problem in the selected data set. The summary of these variables is presented in Tables 9.1 and 9.2, respectively. These variables are classified into demographic, economic, and environmental. Gender, age, and education of household head (HHH) and caste are the demographic variables, and their impacts in selecting these RMPs were hypothesized as positive. Here, gender represents male-headed households, and its impact on RMPs was assumed positive, considering the better access to production resources by men. The age of the HHH represents household experience in seed production, and caste indicates whether the sample households belong to the Brahmin caste. Households that belong to the Brahmin caste were assumed better off in adopting RMPs than those from other castes because of better access to resources. Education in this chapter represents better analytical capability and extension contacts, and those having higher

Table 9.1 Summary of explanatory variables used in the model

Variables	Definition	Mean	Expected sign
Gender	HHH male = 1, 0 = otherwise	0.744 ±0.44	+ve
Age of head of household (HHH)	Age of HHH (years)	46.8±11.43	+/−ve
Education of HHH	Education of HHH (years)	7.96±4.02	+ve
Caste	1= Brahmin, 0 = otherwise	0.62±0.49	+ve
Family labor	Labor force unit (LFU)[a]/household	3.44±0.19	+ve
Land	Operational land (ha)	1.15±0.26	+ve
Livestock	Livestock standard unit (LSU)[b] per household	3.86±2.97	+ve
Physical index	As explained in Table 9.2	1.02±0.58	+ve
Incentive index	As explained in Table 9.2	1.847±0.68	+ve
Technical empowerment index	As explained in Table 9.2	1.433±0.59	+ve
Soil health index	As explained in Table 9.2	0.915±0.61	+ve

1 US$ = NRs 100

[a]LFU is the measurement of labor force, wherein people from 15 to 59 years old regardless of their sex were categorized as 1 person = 1 LFU, but children (10–14 years old) and elderly people (>59 years old) 1 person = 0.5 LFU

[b]LSU is the aggregates of different types of livestock kept at household in standard unit calculated using the following equivalents: 1 adult buffalo = 1 LSU, I immature buffalo = 0.5 LSU, 1 cow = 0.8 LSU, 1 calf = 0.4 LSU, 1 pig = 0.3 LSU, 1 sheep or goat = 0.2 LSU, and 1 poultry or pigeon =0.1 LSU (Khanal and Maharjan 2014)

education are more likely to choose the RMPs (Valandia et al. 2009). Another demographic variable is family labor, and its impact on RMPs was also assumed positive. Seed production in the study area is mainly carried out by family labor, and laborers from outside the households are required only in the critical crop production periods if existing family laborers are not sufficient to accomplish crop husbandry practices (Khanal and Maharjan 2013). Those with more family labor are more likely to adopt RMPs.

These variables contribute on enhancing the economic benefits as well as safeguard seed producers from risk factors (Valandia et al. 2009).

Similarly, all the economic variables—off-farm income, operational land, livestock, infrastructure index, incentive index, technical index, and soil health index—were assumed to have positive impact on adopting these practices.

Table 9.2 Summary of different indices created to use as explanatory variables in model

Types	Specification	PCA weight
(A) **Infrastructural index:** variability explained in component one = 57.04 %		
1. House type	1 if concrete roof (33.8 %), and 0 otherwise	0.5231
2. Irrigation facility	1 if access to irrigation canal (55 %), and 0 otherwise	0.5875
3. Tractor use	1 if use tractor (84.4 %), and 0 otherwise	0.6174
(B) **Incentive index:** variability explained in component one = 50.28 %		
1. Price	1 if higher price for seed than grain (85.5 %), and 0 otherwise	0.4871
2. Training	1 if received training from/through SPOs (78.3 %), and 0 otherwise	0.4107
3. Loan	1 if taken loan from SPO (43.8 %), and 0 otherwise	0.2735
4. Source seed	1 if source seed received through SPOs (92.2 %), and 0 otherwise	0.4627
5. Prestige	1 if realized social prestige (78.8 %), and 0 otherwise	0.4424
6. Profit in seed processing	1 if get economic profit generated by SPOs in marketing (64.8 %), and 0 otherwise	0.3310
(C) **Technical empowerment index:** variability explained in component one = 53.2 %		
1. Knowledge of source seed	1 if heard of any one of the seed categories (breeder/ foundation) (77.22 %), and 0 otherwise	0.4504
2. Household engagement in rogueing	1 if household remove unwanted/diseased plants (86.67 %), 0 otherwise	0.530
3. SPO facilitation in rogueing	1 if SPO members facilitate in rogueing (75 %), and 0 otherwise	0.569
4. Seed lab's inspection in field	1 if field inspection by seed lab (46.6 %), and 0 otherwise	0.437
(D) **Soil health index:** variability explained in component one = 33.4 %		
1. Zero tillage (ZT)	1 if ZT adopted (46.6 %), and 0 otherwise	0.5188
2. Compost improvement	1 if used effective microorganism (20 %), and 0 otherwise	0.3560
3. Green biomass (GB)	1 if GB incorporated (51.6 %), and 0 otherwise	0.3233
4. Seed priming (SP)	1 if SP adopted (60.5 %), and 0 otherwise	0.4726
5. System of rice intensification (SRI)	1 if SRI adopted (12.22 %), and 0 otherwise	0.3728
6. Integrated pest management (IPM)	1 if botanical pesticides adopted, (27.7 %), and 0 otherwise	0.3705

PCA = $\alpha_1 x_1 + \alpha_2 x_2 + \cdots, \alpha_m x_n$, where PC = the principal component, α_1 is weight of variable 1 (value ranging from 1 to n), and $x_1 \ldots x_n$ are the observations

9.3 Results

9.3.1 Risk Incidence, Risk Gap, and Risk Severity

This study shows that 67.22 % of the surveyed households are at profit risk, and a similar level of risk gap (65.23 %) exists among households facing profit risk; however, the risk severity is quite less (42.54 %) than the aforementioned figures.

Table 9.3 Risk incidence and risk gap between different household categories

Variables	Categories	*n*	Risk incidence (%)	Risk gap (%)	Risk severity (%)
Caste	Brahmin	76	66.17	62.6	39.19
	Others	45	67.85	69.6	48.44
Gender of HHH	Male	87	65.41	63.9	40.83
	Female	34	72.34	68.6	47.06
Age of HHH	59 years and younger	109	69.42	64.9	42.12
	Over 59 years	12	52.17	67.4	45.43
Education of HHH	5 years and younger	30	73.17	77.8	60.53
	Over 5 years	91	65.46	61.0	37.21
Cash income	NRs. 50,000 and less	50	85.54	68.3	46.65
	More than NRs. 50,000	71	51.54	60.8	36.97
Operational holding	2 ha and less	113	70.18	66.2	43.82
	More than 2 ha	8	42.1	51.4	26.42
Crop diversification	Not diversified	99	67.78	47.2	22.28
	Diversified	22	64.51	69.2	47.89
Share collection	Collected	68	58.62	57.3	32.83
	Not collected	53	82.81	75.2	56.55
Agricultural training	Attended	90	63.82	60.7	36.84
	Not attended	31	79.48	78.1	61.00

There is also variation in the risk variables between households with different socioeconomic backgrounds (Table 9.3).

For example, risk incidence is higher in households belonging to a non-Brahmin caste as compared to those belonging to the Brahmin caste category. The same situation applies in risk gap and risk severity. A similar result was found in female-headed households compared to those headed by men. The larger risk gap in female-headed households indicates wider disparity in risk in this household category as compared to that of the male-headed group. Similarly, risk incidence is higher in households with younger HHH, but it is just opposite in risk gap and risk severity.

The study shows that the majority of households are smallholders, but risk incidence, risk gap, and risk severity are higher in this category than the counterpart. Similarly, risk is higher in less cash earning households with respect to its incidence, gap, and severity. Those collecting shares in seed producer organizations (SPOs) have less risk incidence, risk gap, and risk severity. Moreover, households receiving training about seed production face less profit risk than those who have not attended training.

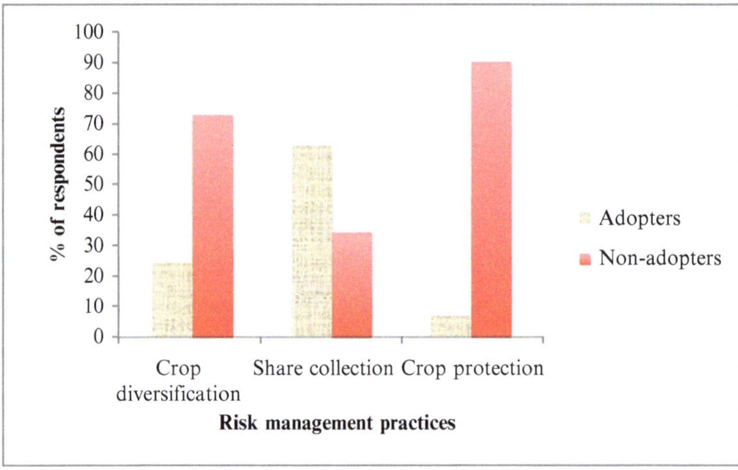

Fig. 9.1 Adoption of risk management practices (From Survey 2011)

9.3.2 Risk Management Practices Adopted by Seed Growers

Seed growers adopt three kinds of RMPs in CSP: crop diversification (25.24 %), share collection (64.44 %), and crop protection (7.8 %) (Fig. 9.1). This result indicates that share collection is the most common practice adopted by seed growers. However, the proportion of share collection adopters is higher in Chitwan as compared to the other two districts. The latter was adopted only in Chitwan District with the facilitation of a development project, and the number of adopters for this practices are very limited. So, the crop protection scheme was dropped from the analysis, and the other two RMPs were considered in the analysis.

9.3.3 Impact of Socioeconomic Variables on Risk Management Practices

The study shows that the selected variables fit the model well, as indicated by the significant Wald test ($p = 0.000$), and it indicates that coefficients of explanatory variables are significantly different from zero (Table 9.4). It is clear from the study that most of the socioeconomic variables have shown the expected sign with dependent variables. However, significant impact was observed in age and education of HHH, caste, family labor, operational land, infrastructure index, incentive index, technical empowerment index, and soil conservation index.

Age has a significant negative impact on share collection, and this shows the higher motivation of younger HHHs in adopting share collection practices in seed production. However, elderly people are more likely to adopt a crop diversification practice. As per the hypothesis, education has a positive impact on share collection

Table 9.4 Coefficient of univariate probit model

Variables	Share collection (1)	Crop diversification (2)
Gender (male = 1)	−0.072 (0.059)	0.209 (0.187)
Age	−0.026 (0.017)**	0.014 (0.0087)
Education	0.055 (0.052)	0.057 (0.042)**
Caste	1.12 (0.875)***	0.951(0.658)
Family labor	−0.050 (0.041)	0.064(0.048)**
Operational land	0.001 (0.001)	−0.0066 (0.006)**
Livestock	0.020 (0.015)	−0.066 (0.047)
Off-farm income	0.238 (0.124)	0.033(0.024)
Infrastructure index	1.391(0.984)***	0.796 (0.478)***
Incentive index	1.535 (0.978)***	0.959 (0.748)***
Technical empowerment index	1.784 (1.458)***	0.705 (0.642)*
Soil health index	−1.077 (−0.985)	0.498 (0.358)***
Constant	2.852	1.799
Model characteristics	$n = 121$, log likelihood = 133.473 ($p = 0.000$), pseudo $R^2 = 0.142$	$n = 121$ log likelihood = 122.473 ($p = 0.002$), pseudo $R^2 = 0.178$

Values in the parentheses indicate marginal impact
*, **, *** indicate significant at 10 %, 5 %, and 1 % level of significance, respectively

practice but its impact on crop diversification is negative, implying that elderly people are less educated but their preference is for adopting crop diversification practices.

Moreover, this study shows that Brahmins are significantly more likely to adopt share collection practices, which might be because of their better contact to extension agencies in accessing production inputs. Similarly, this study shows that households with more family labor are significantly more likely to adopt a crop diversification practice. Moreover, the response of crop diversification is negative toward operational land, which shows the tendency of smallholders toward crop diversification. Those with a higher infrastructural index adopt all three strategies, meaning that those with access to an irrigation facility, good-quality house, and tractor facility are more likely to adopt all three RMPs. The impact of incentive index is also similar to that of infrastructure.

9.4 Discussion

This study analyzed the risk issue in CSP under the rice-wheat cropping system, and the results show that the majority of households are under risk. In the group discussions farmers contended that uncertainty in rainfall has been a challenging issue in recent years in rice and wheat seed production. This conclusion is further justified by the household survey that most households (65 %) considered changing weather patterns (rainfall) as the most important risk in seed production. Other risks in their order of importance are access to source seed (46 %), fertilizer (43 %), and output price (39 %). The rainfall trend in the study areas also supports farmers' opinion because it is quite varied across the years (Fig. 9.2). Rice is grown from June to October (monsoon); therefore, the total amount of rainfall during these months was captured in the analysis. July is the month for rice transplanting, and farmers start seedbed preparation from mid-June. Any deviation in rainfall pattern during July affects rice seed production in quantitative and qualitative terms because it hampers timely transplanting of the young seedlings. Dry weather is expected during rice harvesting time (September/October) from the seed quality perspective because rainfall during this time period caused deterioration in seed quality (seed color and germination). However, the amount of rainfall during the monsoon is in an increasing trend in Chitwan, but it is decreasing by 2.17 mm per month in July, implying that the onset of monsoon rainfall has been shifted later. Similarly, monsoon rainfall is in an increasing trend in Kailali but it is decreasing in October, and this indicates a good weather pattern for rice seed production in

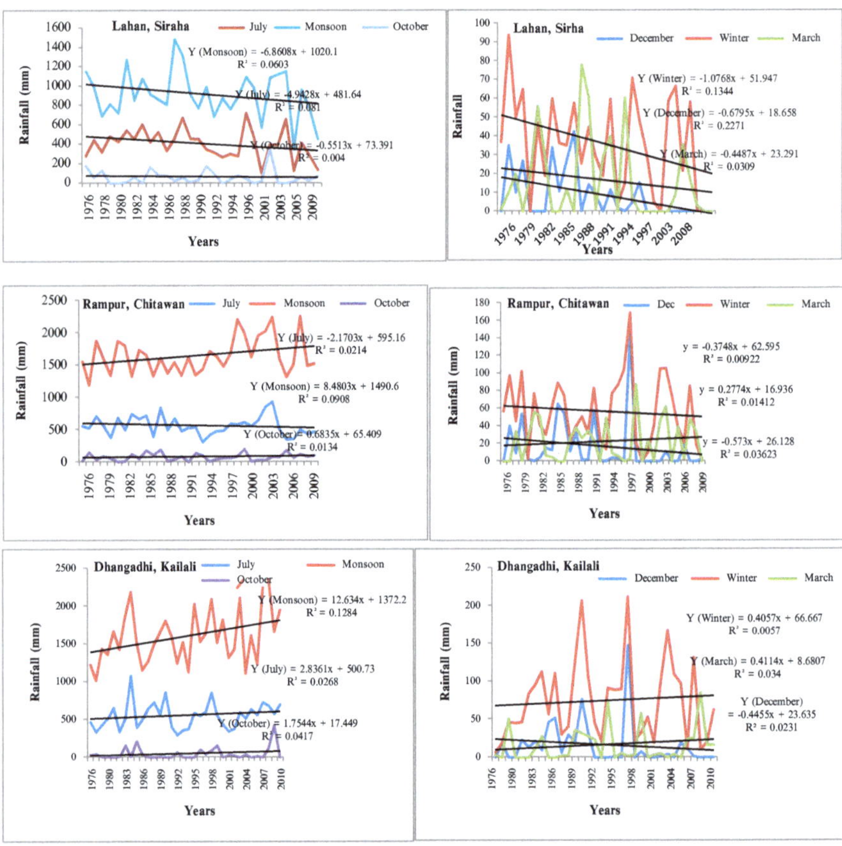

Fig. 9.2 Rainfall trend in summer and winter seasons in the study area (From raw data from Department of Hydrology and Meteorology, Kathmandu, Nepal)

Kailali. However, there is decreasing rainfall trend in Siraha in the entire rice seed production season. Farmers from Chitwan argued that early-maturing rice varieties such as Radha 4 and PR 101 are more prone to damage from rainfall in October.

Wheat is grown from October to March (winter season), and any deviation in rainfall during crown root initiation stage (30–40 days after seed sowing, which falls in December), total winter rainfall, and rainfall during crop harvest (March) would have a negative impact on quantity and quality of wheat seed. As in rice, farmers in Kailali have less risk from rainfall on wheat seed production. However, farmers from Chitwan face risk in wheat seed production because of water shortage during the crown root initiation stage of wheat (from 25 to 35 days after seed sowing in wheat) if availability of an irrigation facility is not considered. The increasing rainfall trend in March would also reduce wheat seed quality.

It is quite obvious that farmers adopt crop diversification strategies to minimize risk from various factors, including climate. It was found that farmers not only diversify crops but also diversify varieties within the crops selected for seed production. The share collection scheme is also related to managing risks from climatic factors. Farmers argue that it might be possible to harvest immature seed crops quite earlier than the recommended date; if farmers wrongly guess the occurrence of rainfall (as there is no system for weather broadcasting in farming communities) there is a possibility of harvesting non-matured seed with reduced quality. But those collecting shares in the organizations feel more accountable toward their organizations and feel more resilient to address the shock (reduced income) because they are more likely to get a loan from the SPOs than their counterparts. This study has also clearly shown that households realizing incentive, with better physical structure (including access to water canal for irrigation) and technical capacity, are more likely to adopt these practices, indicating an avenue for policy implementation. Crop insurance is the most common tool practiced in developed country to address risks in crop production but it has been adopted in the study area. Few farmers adopted this practice with the assistance from the development project and its impact is yet to be analyzed, but farmers believe that households collecting shares are interested in engaging in this practice.

9.5 Conclusion

This chapter analyzed risks in rice and wheat seed production using cross-sectional data. Households performing below the mean level were considered as risky in seed production in the analysis. Seed producers face risks in input and output price as well as crop yield, and the effect of all these variables is reflected in profit. Thus, gross profit made by households in seed production under the rice-wheat cropping system was considered in the risk analysis. The results show that 32.7 % of the surveyed households are at risk, and that there is a 65 % risk gap and 42.54 % risk severity across the households under this category. This study also shows that share collection and crop diversification are the common risk management practices adopted by surveyed households. Moreover, it is clear from the study that households with younger household heads and households associated with the Brahmin caste are more likely to adopt the share collection practice. Similarly, households associated with a less-educated household head and smallholders are more likely to adopt crop diversification practices. Moreover, it is clear from this chapter that infrastructural index, incentive index, and technical index motivate farmers for the adoption of RMPs.

Appendix 9.1: Annual Calendar for Production and Marketing of Rice and Wheat Seed

	Jan	Feb	Mar	Apr	May	Jun	Jul	Aug	Sept	Oct	Nov	Dec
Rice	S	S	S	S	M	P	P	I	I	H	H	C
Wheat	I	I	H	H	C	S	S	S	M	P	P	I

P planting, *I* intercultural operation, *H* harvesting, *C* seed collection from household, *S* storage of seed, *M* seed marketing

References

Almekinders C, Louwaars NP, Bruijin GH (1994) Local seed system and their importance for an improved seed supply in developing countries. Euphytica 78:207–216

Cappelari L, Jenkins SP (2003) Multivariate probit regression using simulated maximum likelihood. Stata J 3(3):278–294

Khanal NP, Maharjan KL (2010) Sustainability of community-based seed production enterprises in Nepal: institutional issues. Nepal Agric Res J 10:33–40

Khanal NP, Maharjan KL (2013) Factors influencing households' behavior in selling rice seed in the market. Agric Food Econ:1–14 http://www.agrifoodecon.com/content/1/1/14

Khanal NP, Maharjan K (2014) Adoption of improved rice varieties in the Tarai region of Nepal. In: Maharjan KL (ed) Communities and livelihood strategies in developing countries. Springer, Tokyo, pp 49–64. doi:10.1007/798-4-431-54774-7-4

Kugbei S (2007) Seed economics: commercial consideration for enterprise management in developing countries. International Center for Agricultural Research in the Dry Areas. Scientific Publishers, Jodhpur

McGuire S, Sperling L (2013) Making seed systems more resilient to stress. Glob Environ Change 23:644–653

MoE (2010) National adaptation program of action (NAPA) to climate change. Nepal climate change portal. Retrieved 12 Sept 2013, from http://www.climatenepal.org.np/main/?p=research&sp=onlinelibrary&opt=detail&id=307

Pradhan KK (2009) Rural finance in Nepal: prospective from cooperative sector. Cooperative Development Board, Kathmandu (Sahakari Sambad Annu Bull)

Shiekh AD, Rahman T, Yates CM (2002) Logit models for identifying the factors that influence the uptake of new no-tillage technologies by farmers in the rice-wheat and the cotton-wheat farming systems of Pakistan's Punjab. Agric Syst 75:79–95

Srinivas T, Bishaw Z, Rizvi J, Niane AA, Manan AR, Amegbeto K (2010) ICARDA's approach in seed delivery: technical performance and sustainability of village-based seed enterprises in Afghanistan. J New Seeds 11(2):138–163

Tabet A (2007) Bayesian inference in the multivariate probit model: estimation of the correlation matrix. Master's thesis. The University of British Columbia, Vancouver

Valandia M, Rejesus RM, Knight TO, Sherick BJ (2009) Factors affecting farmers' utilization of agricultural risk management tools: the case of crop insurance, forward contracting and spreading sales. J Agric Appl Econ 41:107–123

Young GE, Valdez A, Kohn R (2009) Multivariate probit models for conditional claim-types. Insur Math Econ 44:214–228

Chapter 10
Organizational Governance and Its Relationship to Household-Level Economic Indicators: Evidence from Community Seed Production

Abstract Organizations are role-oriented institutions, and in the economic sense they tend to contribute in generating benefit to their members by minimizing costs in input and output marketing. This chapter intends to measure the organizational governance and to assess its linkage with household-level economic indicators. Household data were collected from 180 seed growers and governance data from 12 seed producer organizations of Nepal in 2011. The governance was measured considering organizations' practices in members' participation, business plan development, incentive system, and linkage with service providers. Results show that the organizations are better off in participation and linkage as compared to business plan and incentive system. There is a positive impact of governance indicators on profit efficiency and proportion of rice and wheat seed sold by households in the market. However, the degree of impact of these indicators on the proportion of seed sold is higher as compared to that of profit efficiency. Moreover, organizations with higher educated leaders have better governance indicators. Thus, development of educated leaders would be an important strategy to enhance organization governance and thereby economic benefits at the household level.

Keywords Governance • Incentive • Business plan • Participation • Linkage

10.1 Introduction

Organizing farmers in groups and cooperatives is a popular tool in developing countries for their socioeconomic empowerment (Cook 1995; Acharya 2009). Extension agencies take this approach as a cost effective strategy for delivery of extension services. It is believed that these organizations serve as an innovation platform for members to learn from each other through a self-help approach (Cochrun 1994). Similarly, organizations have potential to enhance economy of scale in marketing of agricultural products, and organized farmers have higher bargaining power in the market chain as compared to individual households. With this assumption, research and development agencies have prioritized the formation and strengthening of agricultural groups or cooperatives in the community seed production (CSP) system. These organizations are termed seed producer

N.P. Khanal, K.L. Maharjan, *Community Seed Production Sustainability in Rice–Wheat Farming*, DOI 10.1007/978-4-431-55474-5_10

organizations (SPOs) in this chapter. In spite of the great potential of SPOs in supplying a diverse varietal choice of food crops in rural communities, the performance of these organizations is not properly understood; some available studies show variation in the performance of these organizations and the reasons for this are not clear (Khanal and Maharjan 2010; Witcombe et al. 2010; Pokhrel 2012). This chapter measures organizational governance and makes its linkage with household-level economic indicators. The economic indicators are profit efficiency (PE) and the proportion of rice and wheat seed sold by households. The PE represents the economic benefits households realize from production stage (from seed sowing to seed sold by households to SPOs). Similarly, the proportion of seed sold by households is a proxy variable to represent the level of the potential benefits of households from seed marketing.

10.2 Conceptual Framework for Measuring Organizational Governance

Organizational governance refers to the instrument that organizations deploy to achieve their intended goal (Hunington 1968). Because seed production in rice and wheat crops is carried out at the household level and seed marketing is handled by SPOs, the governance of SPOs is analyzed putting rice and wheat seed marketing in that context. As already discussed, there might be wide socioeconomic differences among the SPOs members because these organizations were developed to contribute to the socioeconomic condition of farmers residing in the particular geographic area, and this means the criteria to participate in such organizations are residence within a geographic boundary and involved in agricultural activities. So, participants of these organizations are more likely to have heterogeneity in demographic, economic, and institutional resources. This trend might lead to inefficiency of SPOs in marketing because of linkage of these resources with variability, frequency, and economy of scale of SPO output (processed seed). For example, poorer members of SPOs might supply a lesser proportion of their total produced seed to their SPOs as compared to richer members for reasons of food insecurity. Also, being small organizations owned by small farmers, SPOs have to address risks from external factors such as government policy, climate, and market through contingent decision. To address the external factors, SPOs could develop mechanical, adaptive, reactive, or interactive strategies and make contingent decisions (Brinkerhoff and Goldsmith 1990) in line with an organization's efficiency. The governance system contributes in addressing these strategies as it defines a mechanism for maintaining authority, formality, hierarchy, and information flow.

Usually, each SPO forms an executive committee from their members to make decisions in the organization following democratic principles. It is believed that the governance system developed by the executive body will address internal and external challenges faced by the SPO. For example, the incentive system could address the issue of variability, frequency, and economy of scale. Similarly, member participation could also contribute to the efficiency of the organization by enhancing the members' accountability toward their organizations. More informed households would be more loyal and more accountable toward their organization's decisions (White 1984). Organizational governance can be assessed by measuring the performance of the SPO executive body in designing strategies addressing the aforementioned internal and external challenges: participation, incentive system, business plan, and linkage. It is believed that these strategies could also enhance institutional innovations for organizational efficiency in different risk scenarios (Cromwell and Wiggins 1993; Mywish et al. 1999; David 2004; Bishaw and van Gastel 2008).

10.3 Methodology

10.3.1 Study Area and Sampling Technique

This study was carried out in three Tarai districts of Nepal: Siraha, Chitwan, and Kailali, representing eastern, central, and western parts of the country. The Tarai region (70–650 m above mean sea level) is the major food basket of the country, contributing 70 % of the total rice and wheat production. District selection was purposive, but four SPOs with at least 2 years of experience in producing seeds of these crops, and registered in district agricultural development offices, were selected randomly from the available SPOs listed in each district. Then, 15 households from each SPO were chosen for the household survey, making the total sample size 180. Appendix 10.1 depicts the profile of these SPOs. Information related to governance was collected through group discussion with SPO executive committee members and study of existing SPO documents and facilities.

10.3.2 Indicators for Organizational Governance

Four indicators (participation, incentive system, business plan, and linkage) were used to assess organizational governance. However, five subindicators under each of the these indicators were developed based on the existing literature and farmers' experience from the perspective of what makes the SPOs successful to achieve their intended goals. For example, in case of 'participation,' subindicators were developed relative to the vulnerable members to participate and in what activities they need to participate considering the welfare of all the members. The study considers women's participation, poorer members' concerns, members' participation in the annual meeting, and activeness of subcommittee members (technical, financial, and marketing subcommittees).

It was hypothesized that addition of new members to the existing SPOs could enhance social capital and economy of scale in seed marketing. Similarly, the business plan is the key operational document, showing how organizations implement their policies to achieve intended outputs and to minimize risks from internal and external factors. The SPO business plans were analyzed considering the clarity of subcommittee members' roles to implement their activities and the SPO methods in market research, product diversification, quality control, and publicity of seed in the market. The SPOs stated that members could realize incentive in two ways: economic benefit and social benefit (transparency of information). The subindicators reflecting economic benefits include collecting shares (money) in the organization, as it could enhance members' motivation to sell seed in the market through their SPOs, a payment system for executive members based on their workload, and an incentive system to seed growers so they could sell most of the seed their households produce to their organization. Similarly, indicators reflecting transparency in the organization include a system of sharing executive committee decisions with general members and a system for common property management. The common property in this case are materials (e.g., sprayers to manage diseases and pests) that SPOs obtain from development projects. These materials may be utilized for a household's benefit in addition to their common benefit when used at the organizational level. It would be more likely that executive members misuse their power in using these materials in their personal activities if a proper system is not established.

Similarly, SPOs need to maintain good linkage with agriculture research stations to enhance access to source seed, laboratory facilities for testing seed quality, and to access credit as well as training from extension agencies (David 2004). Details of subindicators associated with the aforementioned indicators are summarized in Appendix 10.1.

10.3.3 Measurement and Ranking of Indicators

Each subindicator receives a score from 1 to 4, where 4 represents the best performance (USAID 2005), based on their level of development (Appendix 10.1). The subindicators and their score values were validated in two SPOs not included in the study sample before implementation of the field study. After assigning a score for each subindicator, average scores of the major indicators were calculated. Then, using the average score, major indicators are categorized as low, average, good, and very good. The relationship of these categories and score is as follows:

Score <2.5 = low
Score of 2.5–3.1 = average
Score of 3.2–3.7 = good
Score >3.7 = very good

10.3.4 Measurement of Profit Efficiency in Rice and Wheat Seed Production

Profit efficiency of households in seed production under the rice-wheat system was measured through the stochastic frontier production model developed by Aigner et al. (1977) and Meeusen and van den Broeck (1977). This model is considered better than a deterministic model as it removes the nonsystematic variation in the model and increases the precision of interest variables. Previous scholars have also used this model (Rahman 2003), given as follows:

$$\text{Ln}\, Y_i = \beta_0 + \beta \text{Ln}\, x_i + v_i - \mu_i \tag{10.1}$$

Here, Ln is the logarithm, Y_i is gross profit from seed production under the rice-wheat system (NRs ha^{-1}), β is the vector of parameters to be estimated, and x_i represent inputs. These inputs include labor (NRs per labor force unit, LFU[1]), chemical fertilizers (money spent for chemical fertilizers, NRs/ha), livestock (livestock standard unit, LSU[2], as a proxy indicator to represent amount of animal manure applied in the field), and land (land used in rice seed production, ha). v_i represents the two-tailed error term accounting for random variation in output from factors outside the control of farmers such as measurement errors. Another term, μ_i, represents the error term associated with farm-level inefficiency and is assumed to have zero mean with variance (σ_μ^2) and distributed half normally.

$$\text{PE}_i = Y_i/Y_i^* = f(X_i;\beta)\exp(v_i - \mu_i)/f(X_i;\beta)\exp(v_i) = \exp(-\mu_i) \tag{10.2}$$

where Y_i^* is the maximum possible output, and Y_i, x_i, β, v_i, and μ_i are as explained earlier. PE_i indicates profit efficiency, a measure of the output of the farm relative to the maximum output using the same input vectors. The value of PE_i ranges from 0 to 1. If $\text{PE}_i = 1$, Y_i achieves the maximum value of $f(x_i;\beta)\exp(v_i)$ and $\text{PE}_i < 1$ represents the shortfall of production from the maximum possible production level in the environment characterized by stochastic elements that vary across the farmers.

10.4 Results and Discussion

10.4.1 Overall Performance

In general, SPOs have better performance in participation and linkage as compared to the business plan and incentive system (Fig. 10.1). However, there is quite an amount of variation among these organizations with reference to the aforementioned indicators. SPOs from Chitwan District (Bijbridhi, Pragati, Shreeram, and Unnat) are better in these indicators than those of Siraha and Kailali. Overall,

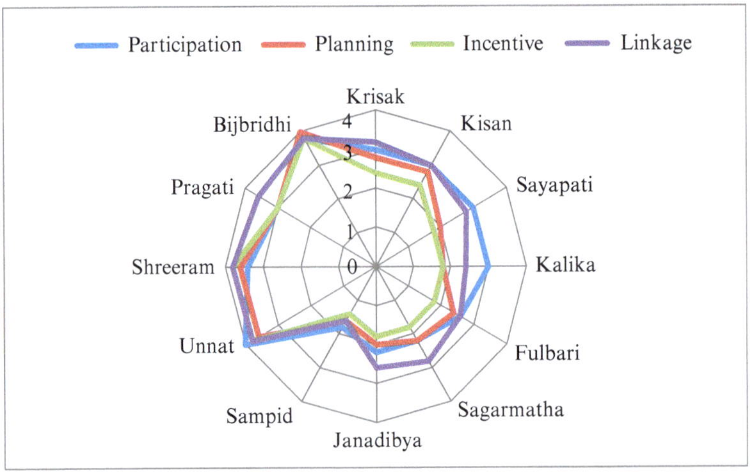

Fig. 10.1 Comparison of governance indicators across the organizations (From Survey 2011)

Sampaid (SPO from Siraha District) showed the least performance with reference to the overall indicators whereas Bijbridhi showed the highest performance.

10.4.2 Performance Relative to Indicators

Participation

Overall, female farmers represent 48 % of the total members of SPOs, and they are also involved in the executive committee, except for three SPOs of Siraha (Table 10.1). Women's involvement in an executive committee means that they could raise their voice in the organizations. But in none of the SPOs were women in the most influential position, that is, chairperson. As mentioned previously, SPOs have heterogeneous members with reference to resource endowment (e.g., land); thus, it might be difficult for poorer households to participate in seed marketing if their organizations do not provide a credit facility or early payment for seed that the households supply to the SPOs. It was found that all SPOs have a policy of prioritizing the poorer people with credit or timely payment for the seed they sell to their organizations. The SPOs of Chitwan and Kailali have adopted the practice of early payment for seed.

However, two SPOs, Unnat and Bijbridhi, have adopted the practice of providing both services (credit facility for implementing seed production activities and early payment of seed for their poorer members). There is no clear-cut written mechanism at SPOs for selecting poorer members; however, executive committee members agreed that they decide which are their poorer members on the basis of land size and annual household cash income.

Table 10.1 Performance of seed production organizations (SPOs) with respect to participation

Districts	SPOs	Subindicators						Remarks
		Women	Poor	General assembly	Subcommittee	Entry of new member	Mean	
Kailali	Krisak	4	3	3	3	2	3	Average
	Kisan	4	3	3	3	2	3	Average
	Sayapatri	4	3	3	3	2	3	Average
	Kalika	4	3	3	3	2	3	Average
Siraha	Fulbari	4	2	3	2	2	2.6	Average
	Sagarmatha	2	2	3	2	2	2.2	Low
	Janadibya	2	2	3	2	2	2.2	Low
	Sampaid	2	2	2	1	2	1.8	Low
Chitwan	Unnat	4	4	4	4	4	4	Very good
	Shreeram	4	3	4	3	3	3.4	Good
	Pragati	3	3	3	3	3	3.0	Average
	Bijbridhi	4	4	3	4	4	3.8	Very good

Source: Survey 2011

All the organizations have the system of holding general assembly on a yearly basis, and this event is supposed to choose new leadership from the members. However, in most of the SPOs, except Shreeram and Unnat, the same people have been in the executive committee since the beginning of their organizational establishment. It was found that in most cases subcommittees have been formed, but they are functioning in only two SPOs (Unnat and Bijbridhi). In most cases there had been no entry of new members since the establishment of the organization, and those who have been added as members after the establishment of SPOs do not have an equal number of shares to those of founder members. For example, in Shreeram founder members have six shares, with one share being equivalent to NRs 5,000, but newcomers have received shares at the rate of three shares per member. However, newly entered members have not been discriminated in Unnat and Bijbridhi.

Business Plan and Its Implementation

All SPOs have drafted their annual business plan, except in Bijbridhi, where there was no detailed information about who should lead what activity (Table 10.2). It was found that SPOs consult with farmers, agrovets, and NGOs before preparing their annual business plans, but the involvement of these actors varies across these organizations. In Kailali (Sayapatri, Janadibya, and Sagarmatha), there was no system of doing any market research, but they produce seed based on the accessibility of source seed from development projects regardless of the types of rice varieties they receive. In Fulbari and Sampaid, they organize a meeting with the local community before preparing the business plan. The organizations from Kailali and Pragati consult with local agrovets and the local community in this process. The study shows that all the SPOs grow both modern and farmers' varieties of rice but only Krisak, Kisan, Unnat, and Bijbridhi sell fertilizer to their members in addition to seed.

Similarly, all the SPOs sell seeds of other crop varieties; however, maize and kidney bean were found only in Chitwan, although wheat is common across the districts. SPOs state that diversifying products help SPOs to minimize their management costs as well as to reduce the necessity of taking an organizational loan.

Only SPOs of Chitwan sell their seed in truthfully labeled bags (including name of crop and variety, germination percentage, weight, seed treated/not treated with pesticides, and name of the producing organization). However, Janadibya, Sampaid, and Sayapatri SPOs sell rice seed without tagging. Among SPOs of Chitwan, Bijbridhi sells more than 70 % of the total rice seed production using proper labeling and bagging.

Table 10.2 Performance of SPOs with respect to business plan

Districts	SPOs	Subindicators						Remarks
		Role clarity	Market research	Product diversification	Quality assurance	Publicity	Mean	
Kailali	Krisak	2	3	4	3	2	2.8	Average
	Kisan	2	3	4	3	2	2.8	Average
	Sayapatri	2	1	3	2	2	2.0	Low
	Kalika	2	1	2	2	2	1.8	Low
Siraha	Fulbari	2	2	3	3	2	2.4	Low
	Sagarmatha	2	1	3	3	2	2.2	Low
	Janadibya	2	1	3	2	2	2.0	Low
	Sampaid	1	2	2	2	1	1.6	Low
Chitwaan	Unnat	3	4	4	4	3	3.6	Good
	Shreeram	3	4	4	4	3	3.6	Good
	Pragati	2	3	3	4	3	3.0	Average
	Bijbridhi	4	4	4	4	4	4.0	Very good

Source: Survey 2011

Incentive System

All SPOs have adopted the practice of collecting cash amounts in their organizations. They call it 'share,' and there is a system whereby profit made by organizations from seed-marketing activities would be distributed to the members/shareholders based on the proportion of share amount they deposited in the organization. However, fewer than half of the members have collected shares in the SPOs of Siraha and in two SPOs of Kailali. Most of the members (more than 75 %) deposit shares in SPOs at Chitwan. Only two SPOs (Unnat and Bijbridhi) distributed the profit generated from seed marketing to their members based on the proportion of their share ownership (Table 10.3). In other cases, the share amount has contributed to increase their organizations' cash reserve.

The second issue in the incentive system is the provision of incentive to the executives who are involved in the organization's management tasks. In six SPOs (four from Siraha and two from Kailali), there was no system of providing incentive to the executives although they are involved in various stages of seed marketing. Similarly, executive members take some resources from the respective SPOs on a consensus basis, especially at the time of major festivals such as Dashain (festival); there is no written rule how much resource is distributed to the executive members when they are involved in the organizations' tasks. However, in case of three SPOs of Chitwan (Unnat, Bijbridhi, and Shreeram) executive members are paid based on their involvement, especially in rogueing (i.e., removal of diseased or unwanted plants/weeds from seed production plots).

It was found that Unnat, Bijbridhi, and Shreeram provide seed and fertilizer in subsidy to their seed growers, but other organizations have not developed such practices.

Transparency of organizations' decisions to their members is considered to be vital in improving cohesion among the members in any organizations. Members who are more informed about their organizations' decisions are more likely to be more flexible toward these decisions and more accountable toward their organizations (White 1984). It was found that SPOs of Chitwan have better performance in record keeping as compared to SPOs from the other two districts. Moreover, SPOs obtain different materials (such as sprayers and grading machines) from development projects. However, only Bijbridhi has adopted the practice of providing these materials to their members for their household activities on a payment basis.

Linkage

The Nepal Agricultural Research Council (NARC) provides source seed to seed producers whether seed production is carried out individually or by a group, but priority is given for farmers engaged in SPOs. Thus, it is easier for farmers to access source seed if they approach NARC through their organizations. It was found that, except SPOs of Siraha, all other SPOs bring source seed when visiting NARC stations (Table 10.4). However, two-way communication has been established only

Table 10.3 Performance of SPOs with respect to incentive

Districts	SPOs	Subindicators					Mean	Remarks
		Share collection	Incentive to executives	Incentive to growers	Information management	Common property		
Kailali	Krisak	3	2	2	2	3	2.4	Low
	Kisan	3	2	2	2	3	2.4	Low
	Sayapatri	2	1	2	2	2	1.8	Low
	Kalika	2	1	2	2	2	1.8	Low
Siraha	Fulbari	2	1	2	2	2	1.8	Low
	Sagarmatha	2	1	2	2	2	1.8	Low
	Janadibya	2	1	2	2	2	1.8	Low
	Sampaid	2	1	2	1	1	1.4	Low
Chitwan	Unnat	4	4	4	4	3	3.8	Very good
	Shreeram	4	4	4	4	3	3.8	Very good
	Pragati	4	3	2	3	3	3.0	Average
	Bijbridhi	4	4	3	4	4	3.8	Very good

Source: Survey 2011

Table 10.4 Performance of SPOs with respect to linkage with service providers

Districts	CBSPOs	Subindicators						Remarks
		Agricultural research	Laboratory	Agricultural extension	Village development committee	Government bank	Mean	
Kailali	Krisak	3	3	4	3	3	3.2	Good
	Kisan	3	3	4	3	2	3.0	Average
	Sayapatri	3	2	3	4	2	2.8	Average
	Kalika	3	2	3	2	2	2.4	Low
Siraha	Fulbari	3	3	3	2	2	2.6	Average
	Sagarmatha	3	3	4	2	2	2.8	Average
	Janadibya	3	3	3	2	2	2.6	Average
	Sampaid	2	2	2	1	1	1.6	Low
Chitwan	Unnat	4	4	4	3	4	3.8	Very good
	Shreeram	4	4	4	3	4	3.8	Very good
	Pragati	4	4	4	2	4	3.6	Good
	Bijbridhi	4	4	4	3	4	3.8	Very good

Source: Survey 2011

in Chitwan, meaning that in Chitwan not only do SPOs visit NARC stations to access source seed but also NARC's professionals visit SPOs in the process of monitoring seed crops in the field. SPOs commented that these NARC visits have been useful to enhance seed quality because farmers get technical advice from these professionals about pests and disease management as well as rogueing.

SPOs were also found to have consulted with seed labs for testing seed quality and with District Agriculture Development Offices (DADOs) to access agricultural training. In these districts, the relationship of SPOs with seed labs and DADOs is similar to that with NARC stations. Moreover, even if the National Seed Policy 2000-envisioned Village Development Committee (VDC) is an important local resource center to support SPOs from the government side, there is poor coordination of SPOs with VDC. Except in SPOs of Sayapatri, where a seed storage house has been built with partial funding from VDC, there is poor communication between VDCs and SPOs. The SPOs of Chitwan have taken a loan from the Nepalese government bank, named the Agriculture Development Bank, which has a mandate to provide loans to the farmers, using their executive members' property (e.g., land, houses) as collateral. The bank's policy on asking collateral while providing loans to SPOs might hamper seed production and marketing because SPOs are run and managed by smallholder farmers. In Siraha and Kailali, SPOs get loans from their members where the interest rate is quite similar to that of the Agricultural Development Bank.

10.4.3 Impact of Governance Indicators on Economic Indicators

Governance indicators have a positive impact on household-level PE and the proportion of seed sold by households in the market. However, the degree of impact of the governance indicators on the proportion of seed sold is higher than on PE (Figs. 10.2, 10.3, 10.4, 10.5). The coefficient for the impact of participation on PE is 9.38, which means that one unit increase in participation tends to increase the PE of household by 9.38 %. However, linkage has the highest impact on PE and the proportion of seed sold (Fig. 10.5); one unit increase in linkage leads to increasing the proportion of a household's seed sold by 23.70 %.

To complement the forgoing analysis, the governance indicators and economic indicators were summarized at the SPO level (Table 10.5). It is clear from the table that SPOs of Chitwan have better economic and governance indicators as compared to those from the other two districts. Moreover, the governance indicators were also compared with characteristics of the SPO leaders (Table 10.6), considering that their leaders' characteristics could be related to the organizations' performance in governance. Although there are 7 to 11 members in the executive committee of the selected SPOs, chairperson and secretary were chosen in the analysis because the

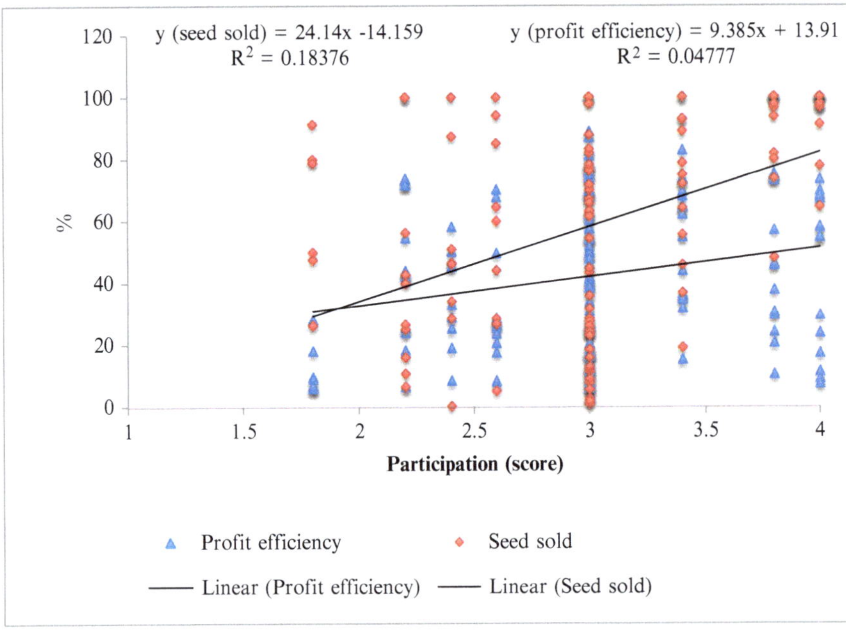

Fig. 10.2 Impact of participation on profit efficiency and proportion of seed sold (From Survey 2011)

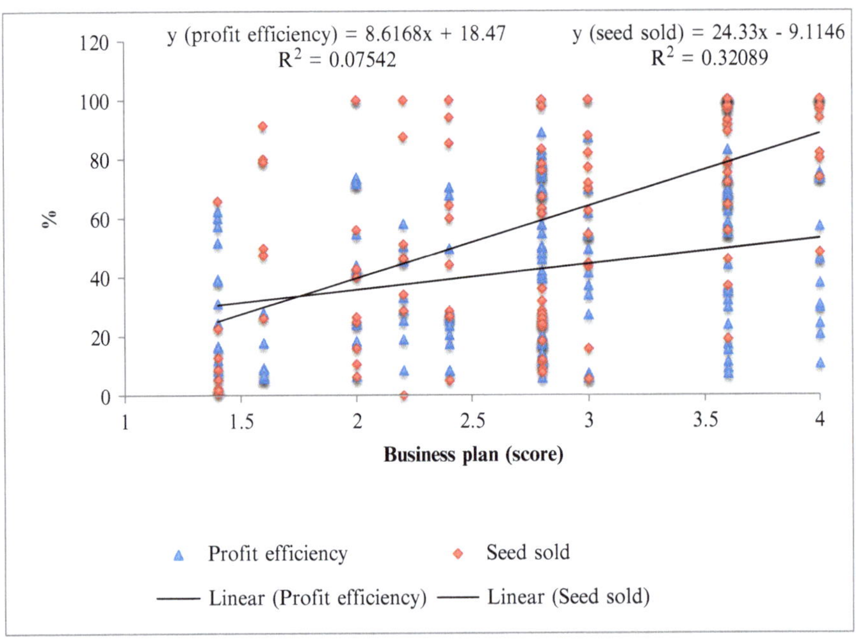

Fig. 10.3 Impact of business plan on profit efficiency and proportion of seed sold (From Survey 2011)

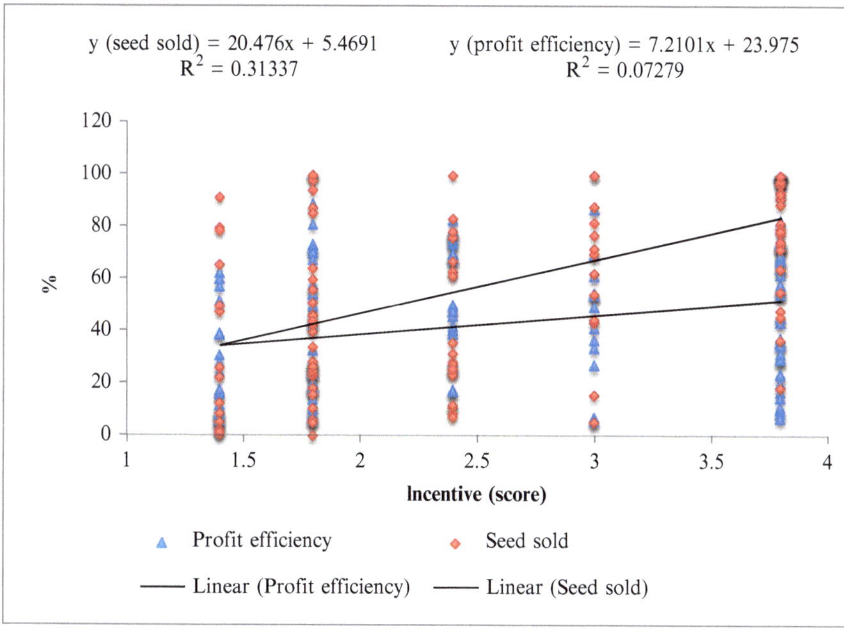

Fig. 10.4 Impact of incentive system on profit efficiency and proportion of seed sold (From Survey 2011)

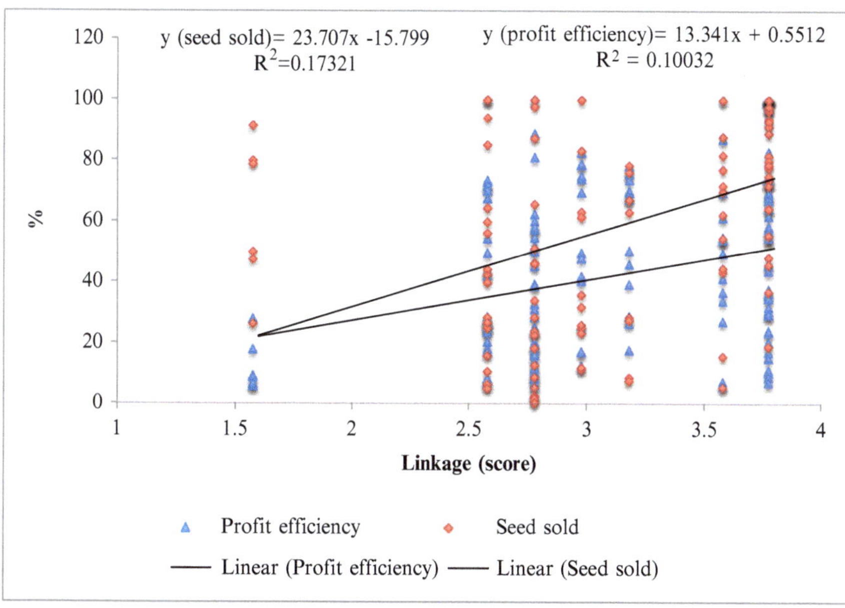

Fig. 10.5 Impact of linkage on profit efficiency and proportion of seed sold (From Survey 2011)

Table 10.5 Household-level governance and economic indicators across SPOs

District	CBSPOs	Participation	Planning	Incentive	Linkage	Profit efficiency (%)	Seed sold (%)
Kailali	Krisak	3.0	2.8	2.4	3.2	56.44	25.52
	Kisan	3.0	2.8	2.4	3.0	50.98	33.85
	Sayapati	3.0	2.0	1.8	2.8	37.78	26.33
	Kalika	3.0	1.8	1.8	2.4	31.25	9.14
Siraha	Fulbari	2.6	2.4	1.8	2.6	29.42	48.68
	Sagarmatha	2.2	2.2	1.8	2.8	29.92	49.30
	Janadibya	2.2	2.0	1.8	2.6	42.03	42.40
	Sampid	1.8	1.6	1.4	1.6	10.78	53.47
Chitwan	Unnat	4.0	3.6	3.8	3.8	42.45	94.44
	Shreeram	3.4	3.6	3.8	3.8	54.21	61.48
	Pragati	3.0	3.0	3.0	3.6	44.75	62.73
	Bijbridhi	3.8	4.0	3.8	3.8	46.24	90.13

Source: Survey 2011

SPOs stated that these positions are most influential in the SPO decision-making process. So, the characteristics (age, years of formal education and training) of these two positions were compared with SPO governance indicators.

Here, age represents experience whereas education and training represent the intellectual ability of the leaders, meaning that SPOs leaders more highly intellectual can have better performance in governance. There is similarity in age of the leaders across SPOs. However, variation exists in education level and leader attendance at business plan training. Leaders' education level is higher in Chitwan as compared to Siraha and Kailali (Table 10.6). There is also a similar trend in average education level of SPO members across the districts (Chitwan, 10.4 years; Kailali, 6.0 years; Siraha, 6.5 years); this means the average education level of the general members reflects the leaders' education in this study. Similarly, SPO leaders from Chitwan district have received business plan training from NGOs, but this was lacking in the other two districts. The attendance on business plan training by SPOs of Chitwan might be a consequence of their higher education level as higher-educated leaders might have better linkage with development projects.

Previous studies have also recognized the importance of education for the better performance of agricultural cooperatives (Acharya 2009; Witcombe et al. 2010) as the leaders having these skills could show better performance in organizational governance. However, Nkhoma (2011) argued that illiterate leaders are more likely to be corrupted and opportunistic, and they turn the organizations toward financial mismanagement and nepotism. These types of leaders might not want to develop a system for proper allocation of incentives in a transparent way.

Similarly, accountability is another aspect affected by low education level. Generally, less-educated leaders are less accountable toward what they are supposed to do. These leaders have a better opportunity to misuse power, such as

Table 10.6 Comparison of governance indicators with their leaders' characteristics

Districts	SPOs	Participation	Planning	Incentive	Linkage	Chairperson			Secretary		
						Age	Edu.	Train.	Age	Edu.	Train.
Kailali	Krisak	3.0	3.0	2.4	3.2	43	7	1, 2, 5	42	10	1, 2, 5
	Kisan	2.8	2.8	2.4	3.0	58	4	1, 2	45	7	1
	Sayapati	2.0	2.0	1.8	2.8	45	10	1	42	10	1, 2
	Kalika	1.8	1.8	1.8	2.4	48	8	1, 2,	49	10	1
Siraha	Fulbari	2.4	2.4	1.8	2.6	66	8	1, 2	35	14	1, 5
	Sagarmatha	2.2	2.2	1.8	2.8	39	14	1, 2, 3, 5	45	10	1
	Janadibya	2.0	2.0	1.8	2.6	35	12	1, 2, 3, 5	29	10	1
	Sampaid	1.6	1.6	1.4	1.6	38	12	1, 2, 3	45	10	1
Chitwan	Unnat	3.6	3.6	3.8	3.8	45	14	1, 2, 3, 4	50	12	1, 2, 3, 4
	Shreeram	3.6	3.6	3.8	3.8	60	12	1, 2, 3, 4	60	12	1
	Pragati	3.0	3.0	3.0	3.6	67	12	1, 3, 5	32	12	1, 5,4
	Pithuwa	4.0	4.0	3.8	3.8	70	11	1, 2, 3, 4, 5	47	14	1, 2, 3, 4, 5

Source: Survey 2011
Participa. participation, *Edu.* education (i.e., formal schooling years), *Train.* training (*1* seed production, *2* marketing, *3* leadership, *4* business plan, *5* account)

diverting activities in accordance to their own priorities without doing proper consultation with other members or designing activities in the interest of political parties (Chriwa et al. 2005). It is clear from the study that three SPOs especially, namely Bijbridhi, Unnat, and Shreeram, are better in both economic and governance indicators.

These three organizations were also promoted by development projects but leaders of these organizations were school teachers who are more highly educated than other members. Being local teachers, they had the capacity to motivate farmers to organize in groups/cooperatives, developed planning and incentive systems, and could make linkage with development projects to access resources.

When these organizations implemented the share collection policy, some members dropped out of the organizations because they were not confident about the safety of their investment. But after a few years (especially in Unnat and Bijbridhi), some of those farmers who had dropped out rejoined the same organizations as they viewed the SPO's progress. Thus, better-performing SPOs have tried to generate and promote innovation in their development pathway, which is driven by efficiency gain, and this phenomenon is similar to what Morris et al. (1998) used to discuss about the evolution of the maize seed industry. It is quite difficult to expect that all SPOs will go through the same model and achieve the same level of performance, but the process of their transformation might be useful to those working in community seed production. We discuss this matter in the next section.

10.5 A Case of Institutional Innovation in Bijbridhi

10.5.1 Motivation for Seed Production

The Bijbridhi SPO was evolved from a group named Pithuwa Biu Utpadak Krisak Samuha (PBUKS), which was formed in August 1994 by nine farmers. These farmers were involved as contract seed producers for Agricultural Input Corporations (AIC), the government corporation responsible for supplying agricultural inputs in Nepal at that time. While engaging in seed production, farmers realized that producing seed would be more profitable than producing grains of cereal crops. Farmers also thought that the AIC was paying them a low rate for their seed, with the justification that seed produced by these farmers was of poor quality (low physical and genetic purity). Similarly, seed produced by farmers was scattered and low in volume, which increased the transportation costs of AIC. Even if they had a contract with AIC, there was no certainty that AIC would buy the total produced seed from the farmers every year as per the contract agreement. For these reasons, farmers decided to form their own organization to produce and sell seed in the market. These farmers also received motivation from NGOs and DADOs to take part in seed production in an organized way.

10.5.2 Functioning of the Group

The group was formed with the objective of improving its members' socioeconomic conditions. This objective was not set by the farmers themselves but by development projects implemented in the area with the poverty reduction motive. After setting up the objective, the group was legally registered in DADO of Chitwan in August 1996. Then, all the members started producing improved seeds of rice, wheat, maize, lentil, and kidney bean, collecting source seed from NARC stations. Looking at the benefits acquired by these farmers from seed production, neighboring farmers, who were growing the aforementioned crops as grain, approached the group for membership. The group decided to increase its members to enhance its economy of scale, and by 2001 the members in the group increased to 61. The newly entered members had to pay a membership fee at the rate of NRs 50 per member (later it was increased to NRs 100). The major attraction of new members to enter into the group was that they would receive extension facilities from DADO (technical training, exposure visits), agricultural research stations (source seed, training), seed laboratory (seed testing facility), and NGOs (training, visits).

The group formed an executive committee from their members wherein the members were selected democratically from the general assembly (annual meeting), and the committee had a 2-year tenure. There were three subcommittees: technical, marketing, and finance. The technical subcommittee had to be involved in the selection of seed grower farmers, maintenance of isolation distance of seed plots, removal of diseased and off-type plants, seed quality inspection at threshing, and pesticide application. Seed quality indicators considered at threshing were diseased seed, inert materials, and seed color. Seed growers agreed that seed color is the most important seed quality indicator for seed consumers (local farmers) as this determines whether the seed has been harvested from a properly matured crop and whether it is infected with disease. Seed crops planted late or soaked by heavy rain during harvesting could not produce a bright shining color on seed coat.

The group also started lending to their group members using the funds collected at the organizational level from various sources. As already discussed, the most common sources were the group membership fees and commission/benefits received from seed sale. For example, members had to pay NRs 2 per kilogram of seed sold to the group fund. Similarly, money collected from other sources, such as fees charged for visitors, and donations given to the group by government and nongovernment agencies was also included in the group fund.

10.5.3 Challenges in the Group

In spite of the concerted effort from various organizations for enhancing the group capacity, the group savings could not be substantially increased (NRs 15,000 in

2005 as compared to NRs 12,000 in 2000). The executive members argued that the major reason for lower group performance was that the group became completely dependent on extension agencies for accessing source seed and selling their seed. Extension agencies bought most of the seed SPO produced for their project, even paying a comparatively higher price than that in the market. As a result, in some years, when projects did not buy seed, farmers could not sell seed as per their plan as the price set by the group was quite a bit higher than that for seed available in the market through other, different means, because the group set the price with reference to what development projects had paid for their seed in previous years. Also, cases of conflict started increasing among the members because of the organizations' inability to develop physical resources for seed processing and marketing, low group savings, and the inability of SPOs to address the concerns of the poorer members.

10.5.4 Conversion of Group to Producers' Seed Company

In 2005, some members of the group, those involved in technical and marketing subcommittees decided to form a producer seed company. These people had already gathered information about seed demand and established good linkage with research and development agencies. In a group discussion, executive committee members of the organization argued that although the proposal for setting up a company was discussed in the SPO's meeting only 16 members agreed on the proposal, and they formally registered the seed company in 2006 in accordance with the Nepalese Company Act 2004. All the members deposited shares amounting to NRs 4,000–10,000, and they bought a seed grader machine and built a seed storage house (Fig. 10.6) using funds collected from share amounts from members and grants they received from development agencies such as DADO (NRs 25,000) and District Development Committee (NRs 20,000). Similarly, responsibility was given to the executive committee members of the organization based on their competency and commitment. For example, the organizational manager was selected from the members with bachelor-level education in commerce, paying his monthly salary. The manager is also the secretary of the organization, and the executive committee has delegated full authority to him to implement activities authorized by the committee. In a meeting, the manager of Bijbridhi argued that he collects information regarding amount of rice seed available with competitors and associated price routinely from the market by telephone calls. He mentioned that collecting these types of information is important to design strategies for addressing uncertainty of seed price.

The company realized that the seed produced only by the existing members would not be sufficient to cover their management cost. So, they contracted with 300 growers (with about 300 ha) residing in six VDCs (Pithuwa, Jutpani, Chainpur, Kathar, Padampur, and Shaktikhore), and one municipality (Ratnanagar). They also increased the number of crops and their varieties in seed production and marketing,

Fig. 10.6 Storage house of seed producer company (From authors' photo bank)

and their volume of seed sold started increasing over the years (Fig. 10.7). The majority of seeds of these varieties (>75 %) are sold in labeled bags with the brand name *Kisan Ko Biu*, meaning farmers' seed, with different packaging sizes, and the type of packaging materials was also based on consumer demand.

Moreover, the company used to sell the majority of its seed to NGOs, DADOs, and local farmers. In 2010, the company sold 557.3 t of seed where the share of rice seed was 45 %. There is increasing trend of rice seed sold by the organization, most of which is sold to agrovets (>90 %), and the remaining to the local farmers. The company record shows that its seed has been disseminated to 35 districts of Nepal. The company has developed an incentive system to seed producers, executive members, and has good linkage with government agencies and NGOs. They produce both local varieties and improved crop varieties to address the consumer demand. The members of these groups also participate in various meetings related to agriculture and the seed system in the country.

10.5.5 Relationship Between Group and Company

The Company has been able to maintain good relationships with PBUK; 25 members of this group are also the shareholders in the company. The non-shareholders of the company, especially the poorer members of the group, are also benefited from the company as they could access quality seed from this organization. Some of the group members sell seed to the company. Similarly, the company provides local farmers with seed, chemical fertilizer, and bio-fertilizers, even with a loan if needed. Moreover, the company is also benefited from the group as it takes a loan

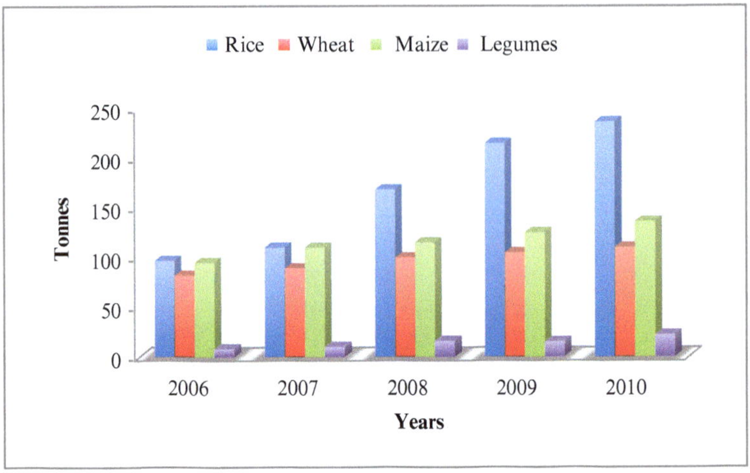

Fig. 10.7 Rice seed sold by Bijbridhi over the years (From Survey 2011)

from PBUK as the members are still continuing monthly saving from the beginning of its organizational establishment. The committee members claimed that PBUK provided the foundation for the establishment of Bijbridhi because those members involved in PBUK internalized the benefits from seed production and marketing activities. As a result, they developed confidence in seed management activities and opened a private company using their own investment.

10.5.6 Reasons for Success of the Group

Among SPOs surveyed in the study, Bijbridhi was found most successful as it has covered the marketing cost (Table 10.5) and provided additional benefits to the members. Also, it has captured the issues of both small and large farmers by including both farmers' variety and modern varieties in seed production plans. Moreover, it has developed strategies to address the risks that create organizational inefficiency in seed marketing by designing governance strategies. There might be various factors contributing to the success of this organization but one clear-cut difference between Bijbridhi SPO and other poor performing SPOs (such as Pragati) is leadership. The group (PBUK) was led by high-school teachers; some other primary school teachers and retired staff from agricultural extension agencies were included as members in the group. The educated leaders allowed the group to become able to develop an incentive system in the organization, and developed seed-processing structures (grading machine, seed storage, and threshing floor) through the public–private partnership approach. Even after changing the organization structure from group to company, the educated and experienced people were selected for the executive committee. Similarly, subcommittee members were also chosen based on their skills and commitments. These leaders also were able to make connection with development projects working in the district to access training on various aspects of seed management. As in the Bijbridhi SPO, the importance of educated leaders is also clear in other SPOs of Chitwan District (Sect. 10.4.3), but this does not mean that leadership is the only factor for the good performance of SPO. Education of seed consumers and accessibility of agricultural inputs such as an irrigation facility, source seed, and an extension facility are better in Chitwan District.

10.5.7 Challenges for the Company

Key challenges faced by Bijbridhi are uncertainty to access source seed as per their demand, an irrigation facility, and risks from rainfall variability. For instance, the organization had requested 10 t source seed of rice from the NARC station, but it supplied only 50 % of the demanded amount in 2010, justifying this by the availability of limited seed quantity at its store. In many cases, the source seed was of poor quality in terms of germination and genetic purity. Unpredictable rainfall has been a key concern of seed production in this SPO. In 2011, the organization failed to collect 20 t of rice and 25 t of wheat, which accounts for 10 and 15 % of the total estimated rice and wheat seed production in the organization.

10.6 Conclusion

This chapter measured the governance of SPOs with respect to participation, business plan, incentive system, and linkage. These governance indicators were regressed with household-level economic indicators (profit efficiency and proportion of seed sold by seed producers in the market) to see their relationship. The governance indicators were measured from organizational policies and practices, whereas economic indicators were based on household-level data. Results show that, in general, SPOs have better performance in participation and linkage as compared to the incentive system and business plan. In most SPO leaders do not get economic benefits directly from SPOs, but the social benefits such as social prestige and increased networking with different service providers motivates them to be in leadership. It is necessary to consider these social benefits in evaluating the performance of SPOs because the objective of these organizations is not limited to contributing to household profit from seed production but also contributes to their socioeconomic development.

Moreover, it was found that most of the organizations have a policy to support poorer members by the provision of credit and timely payment for their seed, but this has been adopted only in a few cases. There is a positive impact of governance indicators on households' economic indicators, and SPOs with higher educated leaders have better governance indicators, possibly because more highly educated leaders are more able to access services from service providers, to develop the incentive system for enhancing members' accountability toward SPOs of members in seed production and marketing, and to prepare an appropriate seed production and marketing plan. This input suggests to development agencies how they could improve performance of SPOs through education, but this does not mean that provision of an educational program alone is sufficient to strengthen community seed production. Lack of source seed, poor soil fertility, risks from climate stress, and poor seed processing and irrigation facilities are the common concerns for the sustainability of this system.

It is also clear from the analysis that CSP serves as a platform to empower farmers in seed production and marketing. Once it becomes mature, some of the members could set up new local corporations. This transformation happens in those areas where farmers have better access to physical and institutional facilities. The poor members might not join in the newly formed entity because of their resource constraints but they will benefit from the availability of seed supplied by the newly formed entity at a reasonable price at the local level.

Appendix 10.1: Indicators and Scores Used to Assess the Capacity of SPOs

1. *Participation*

	Scores			
Subindicators	1	2	3	4
1.1 Women participation	<10 % women members in the organization	11–25 % women members in the organization	26–50 % women members in the organization	Women in the executive committee
1.2 Participation of poor (support strategies to poor)	No written rule	Written rule exists but not operationalized	Special consideration for poor in credit or timely payment	Special consideration for poor in payment and credit both
1.3 General assembly (annual meeting of CBSPOs)	Never held	Held but not regular	Regular but same members in the executive committee from the beginning	Held regularly, and some members changed
1.4 Subcommittee	Not formed	Formed but not functional (no meeting within a year)	At least one subcommittee functional (≥2 meetings in a year)	At least two committees functional

(continued)

Subindicators	Scores 1	2	3	4
1.5 Entry of new members	No system for entry of new members (only founder members exist)	System exists but no members entered in the organization	New people entered in the organization without equal share	New people entered in the organization with the provision of equal share

2. *Business plan and its implementation*

Sub-indicators	Scores 1	2	3	4
2.1 Role clarity in the business plan	Not available	Available in draft form but operational plan not developed	Operational plan developed but roles not specified	Detail operational plan developed and roles specified

Business plan. . ..

Subindicators	Scores 1	2	3	4
2.2 Market research	Consultation is not done with stakeholders	Consult with local farmers	Consultation local farmers and local agrovets	Consultation with farmers, local and distant agrovets
2.3 Product diversification	Seed production of only one crop	Seed production of two or more crops	Two or more crops and inclusion of local varieties	Sell two or more crops seed and other inputs
2.4 Seed quality assurance measures	Simple bagging but no tagging	Seed packaging in branded bags but no tagging	Seed packaging in branded bags, use of tagging for <50 % seed	Seed packaging in branded bags for >50 % seed
2.5 Publicity of products	No publicity	Sending letter to organizations	Sending letter and publicity in agriculture fairs	Publicity through FM radio

3. *Incentive system*

Subindicators	Score 1	2	3	4
3.1 Share collection from members in the organization	No system of collecting share	Less than 50 % of the members	50–75 % of the members	More than 75 % of the members

(continued)

Subindicators	Score 1	2	3	4
3.2 Incentive to executives	All voluntarily	Occasional basis only to chairperson	Occasional basis both chairperson and executives	Defined norms to pay chairperson and executives
3.3 Incentives to growers	No system for providing incentive to seed growers	Technical facilitation or subsidy on fertilizer/seed exists	Technical facilitation and subsidy exist but not crop insurance	Technical facilitation, subsidy and crop insurance

Incentive system...

Subindicators	Scores 1	2	3	4
3.4 Information management	Written documents do not exist	Very raw, unclear and poor record-keeping system	Draft type of simple record-keeping system	Good record-keeping system using ledger books
3.5 Common property management	No system for the use of common property	System exists but not in function	Mobilized based on rotation	Mobilized based on payment to the organization

4. *Linkage with service providers*

Subindicators	Scores 1	2	3	4
4.1 Linkage of CBSPOs with agricultural stations (NARC) for source seeds	No linkage	Poor linkage with some communication	Visit to NARC station and source seed received	Two-way visits and source seed received
4.2 Linkage of CBSPOs with seed testing laboratory	No linkage	Poor linkage with some communication	Visit to seed laboratory and services received	Two-way communication between seed laboratory and CBSPOs
4.3 Linkage of CBSPOs with VDC	No linkage	Poor linkage with some communication	Visit VDCs and formal communication exist	Resource tapping from the organization
4.4 Linkage of CBSPOs government bank	No linkage	Poor linkage with some communication	Visit bank and formal communication exist	Resource tapping from the organization
4.5 Linkage of CBSPOs with DADOs	No linkage	Poor linkage with some communication	Visit DADOs and formal communication exist	Good linkage (received training or other sources)

References

Acharya BM (2009) Is cooperative really a democratic organization? Lessons learnt from the selected cooperative acts of the world. Cooperative Development Board, Kathmandu (Sahakari Sambad Annu Bull C)

Aigner D, Lovell CAK, Schmidtip (1977) Formulation and estimation of stochastic frontier production model. J Econ 6:21–37

Bishaw Z, van Gastel AJG (2008) ICARDA's seed delivery approach in less favorable areas through village-based seed enterprises: conceptual and organizational issues. J New Seeds 9(1):68–88

Brinkerhoff DW, Goldsmith A (1990) Institutional sustainability in agriculture and rural development. Praeger, New York

Chriwa E, Dorward A, Kachule R, Kumwenda L, Kydd J, Pool N (2005) Farmer organizations for market access: principles for policy and practice. Department of Agricultural Science, Imperial College, London

Cochrun SE (1994) Understanding and enhancing neighborhood sense of community. J Plan Lit 9:92–99

Cook L (1995) The future of US agricultural cooperatives: a neo-institutional approach. Accessed 10 Apr from http://web.missouri.edu/~cookml/publicat.htm-35k

Cromwell E, Wiggins S (1993) Sowing beyond the state: NGO and seed supply in developing countries. Retrieved April 1, 2011, from http://linkinghub.elsevier.com/retrieve/pii/0305750X9400133J

David S (2004) Farmer seed enterprises: a sustainable approach to seed delivery? J Agric HumValues 21:387–397

Hunington SP (1968) Political order in changing society. Yale University Press, New Haven

Khanal NP, Maharjan KL (2010) Sustainability of community-based seed production enterprises in Nepal: institutional issues. Nepal Agric Res J 10:33–40

Meeusen W, van den Broeck J (1977) Efficiency estimates from Cobb–Douglas production function with composed error. Int Econ Rev 18:435–444

Morris ML, Rusike J, Smale M (1998) Maize seed industries: a conceptual framework. In: Morris ML (ed) Maize seed industries in developing countries. Lynne Rienner, Boulder

Mywish M, Julie H, Ducan B (1999) Increasing seed system efficiency in Africa: concepts, strategies and issues. Retrieved March 15, 2011, from https://ideas.repec.org/p/ags/midiwp/54578.html-12k

Nkhoma AT (2011) Factors affecting sustainability of agricultural cooperatives: lessons from Malawii. Master's thesis. Agricultural Commerce of Massey University, New Zealand

Pokhrel S (2012) Role of DISSPRO and CBSPs in current seed supply situation in Nepal. J Agric Environ 13:53–59

Rahman S (2003) Profit efficiency among Bangladeshi rice farmers. Food Policy 28:487–503. Retrieved February 25, 2012, from http://www.sciencedirect.com/science/article/pii/S0306919203000745

USAID (2005) CBO/FBO capacity analysis: a tool for assessing and building capacities for high quality responses to HIV/AIDS. Retrieved September 10, 2013, Maryland from http://www.medicalteams.org/docs/learning-zone/core_capacity_analysis.pdf?sfvrsn=0

White LG (1984) Lessons in development project management. University of Maryland, College Park, Maryland

Witcombe JR, Devkota KP, Josh KD (2010) Linking community-based seed producers to markets for a sustainable seed supply system. J Exp Agric 46(4):425–437

Chapter 11
Institutionalization of Community Seed Production

Abstract Community seed production initiatives are being implemented in the form of projects, and lessons from these projects should be institutionalized in the programs and policies of corresponding stakeholders for enhancing their impacts for food security of the people. These stakeholders might not be able to adopt the project outputs quickly as technology adoption is a gradual process, so stakeholders need to innovate ideas to promote/adapt lessons generated by the projects in different contexts. Some of the areas for innovation include participatory approach, leadership, education, entrepreneurship skills, source seed production, and its distribution.

Keywords Institutionalization • Dynamic leadership • Participatory approach • Local capacity building • Education

11.1 Introduction

Institutionalization is a process through which new ideas and practices are introduced, accepted, and used by individuals and organizations so that these ideas and practices become part of their norms in their actions. Institutionalization involves changes in stakeholders' behavior and development in response to internal and external stimuli (Ejigu and Waters-Bayer 2005). The concept of this process came in the community seed production (CSP) system to mainstream the lessons generated in the process of its implementation into the stakeholder's programs and policies. Mainstreaming the lessons is important because the CSP intends to catalyze innovations in formal and informal seed systems to strengthen the integrated seed system (ISS) in farming communities. The ISS could address the diverse ranges of crop varieties demanded in rural farm communities and the objectives of the organizations that advocate seed from different perspectives: means of conserving local biodiversity, food security, source of livelihoods, and source of rural entrepreneurship and income. The first three perspectives are development oriented and highlight the values of seed from the social and environmental perspective. However, the fourth is economically focused and discusses how private seed companies emerge and their impacts on food security and income of people. Both these perspectives are important, and how to harmonize both these

N.P. Khanal, K.L. Maharjan, *Community Seed Production Sustainability in Rice–Wheat Farming*, DOI 10.1007/978-4-431-55474-5_11

perspectives in programs and policies of organizations dealing with seed issues is a great concern.

The second value of institutionalizing CSP is to promote accountability of development agencies in delivering extension facilities even after project phase-out. There is a growing realization that organizing seed production demonstrations in some locations and provision of some short-term training (as adopted by most of the development projects) to seed producers would not automatically change their attitudes and behavior to address efficiency and equity issues in seed production and marketing. Rather, it raises concerns for the continuity of such behaviors beyond the project framework. It demands follow-up support from government agencies even after the termination of support from development projects. Ironically, it has been learned that government agencies would not support the outcomes of donor-funded projects through their regular programs because of their resource constraints and lack of policy framework for the documentation and institutionalization of project outcomes (Bishaw and van Gastel 2008; Srinivas et al. 2010).

The institutionalization of CSP is considered important not only for the continuation of project outcomes in the future but also to enhance their impacts in the wider mass by scaling out project lessons through programs of extension agencies (FAO 1998). The integration also contributes in communities' resilience to climate stress (Bishaw and van Gastel 2008; Srinivas et al. 2010; Witcombe et al. 2010; McGuire and Sperling 2011, 2013). This chapter summarizes key lessons learned while analyzing the sustainability of CSP in the preceding analytical chapters (Chaps. 4, 5, 6, 7, 8, 9, and 10), and discusses the process of institutionalizing those lessons.

11.2 Lessons to Be Institutionalized

As discussed in Chap. 3, the major quest of sustainability discussed in this book is how seed producers realize benefits and how these benefits can continue in the future. The benefits can be measured from farmers' performance in utilizing resources (technical efficiency, profit efficiency), environmental performance (farmers' behavior in soil conservation practices), and social performance (organizational governance). It is seen from the analysis that households with higher access to resources are benefited more in CSP although this system is targeted for resource-poor farmers. It raises a critical question about how the benefits of resource-poorer households could be enhanced. Here, the benefit is a relative term, and it would be worthwhile to compare the benefits among the seed producers involved in the same seed producer group/cooperative. Heterogeneity of members with respect to resources, education, and orientation is a common phenomenon in rural seed areas. Therefore, it might not be practicable to consider members' homogeneity with respect to the foregoing characteristics in seed producer groups/cooperatives operating in the initial phase of seed industry development. The key consideration here is how the heterogeneous members in the organization work together for their economic and social benefit. The realization of benefit is the main driver to change seed production from the subsistence level to the commercial stage. For this transformation, it is important to create innovations in the following aspects.

1. Leadership has a key role in enhancing the performance of CSP. Good leaders could effectively address the conflict of interests among members and develop external influences (e.g., politics, government policies) in their favor. The conflict of interests among the members could be harmonized by developing an incentive system, business plan, transparency, and members' participation in the decision-making process. Education contributes in developing dynamic leaders who are better off in designing these strategies as compared to those who are not educated.
2. Entrepreneurship is a key to enhance seed producer performance in seed production and marketing because it enhances farmers' willingness to develop, organize, and manage a business venture along with its risks to achieve the objective of the organization. The concern of entrepreneurship is related to how seed producers increase their group capital, including a seed-processing facility, how to address credit constraints in the SPOs, how to maintain cash flow in a year, what kinds of varieties or complementary activities could be integrated in CSP to enhance economy of scale of SPOs, and so on. Group saving contributes to the SPO capital reserve, and it also makes the member more accountable toward their organization as the benefits collected from seed marketing are distributed among the members based on the amount of money they collect in the group fund.
3. Incentives realized by seed producers are not necessarily economic in nature, but they could realize social benefits such as technical empowerment, social prestige, and networking with different service providers. These benefits explain why executives of seed producers work voluntarily in the early phase of their organizational development. Thus, it is important to integrate social benefits while evaluating the performance of CSP.
4. Maintaining soil health contributes to sustainability of community seed production because it determines current benefits (production efficiency) and potentials for future benefits. The adoption of soil conservation practices (animal manure, zero tillage, green manure, system of rice intensification, and improved composting) provides the basis for harmonizing current and future benefits in seed production.
5. Seed producer groups and cooperatives serve as a platform to enhance the socioeconomic empowerment of their members. These structures are less formal than private companies but accommodate even the very poor farmers. The organizations transform from less formal to more formal, and efficiency gain by the member catalyzes their transformation. The pace of transformation is faster in SPOs with resource-rich members, suggesting development of two types of strategies by development agencies in strengthening CSP. First, SPOs located in a favorable environment (better irrigation facility, road network, and communication facility) should be promoted from the perspective of

transforming them into local seed companies. In the transformation process, the whole organizations (groups/cooperatives) might change to companies, or a few resourceful and empowered farmers may set up new companies. It is more likely that poorer members could also directly join in the company if they have access to a credit facility. Even if they could not join, resource-poor farmers are benefited from the newly established local companies as they could access improved seed and associated crop production technologies at the local level in a reasonable price.

Second, development projects should consider SPOs located in environments less favorable from the development perspective. Seed producers/members of these organizations should be made aware of the benefits of using improved seed and mechanisms to access source seed from government organizations or seed companies. The SPOs might not sell a large volume of seed outside the organization but could take the membership of well-functioning seed companies located in the accessible area. In this case, SPOs could focus their activities in seed production whereas seed companies facilitate seed marketing.

11.3 Future Directions in Institutionalizing Community Seed Production

The community seed production system, as an integrated system, catalyzes innovations in formal and informal systems to achieve the broad objective of enhancing food security through seed security. For this, stakeholders should integrate lessons of CSP implementation into their working modalities. Knowledge integration is a continuous process and it might be changing over time. However, it should address the efficiency of the seed production system and development constraints such as empowerment, transparency, and inclusion. This section discusses the innovations to be made by formal and informal stakeholders with reference to the problems that hinder the strengthening of the CSP system. Stakeholders associated with the formal system are national agricultural research and extension agencies, regulatory bodies, and local government; and those included under the informal system are seed producer organizations (SPOs) and nongovernmental organizations (NGOs).

11.3.1 Formal Stakeholders

National agricultural research centers have the major role in supplying source seed of crop varieties to seed producers. However, varieties included for seed production in these organizations are not appropriate for smallholder farmers residing in marginal environments. To address this issue, it is important to increase

Table 11.1 Innovations with reference to problems across the formal stakeholders

Problems	Innovations
National agricultural research system	
Lack of source seed of varieties preferred by farmers	Effective models for actor participation (e.g., participatory plant breeding)
	Efficient technique for the registration of crop varieties
Low volume of source seed available	Cost–benefit analysis of source seed multiplication channels
	Development of source seed production centers at local levels
Government extension	
Focus on technical training	Methods to integrate entrepreneurship in different levels of training
Poor access in marginal areas	Mechanisms to integrate seed delivery techniques with local NGOs
Regulatory body	
Limited capacity and centrally located	Promote truthful labeling in community seed production (CSP)
	Incentive mechanism is needed to promote seed lab at local level
	Strategies to stop/monitor malpractice in seed production, processing, and distribution
Local government	
Poor linkage with CSP	Documentation of seed producer groups and cooperatives
	Strategies to motivate local people to establish source seed production centers/seed lab
	Appropriate business models to foster public–private partnership

participation of farmers in variety development phase, which can be done by adopting the participatory plant breeding (PPB) approach in the process of crop varietal development (Table 11.1). Participatory variety selection (PVS) and client-oriented breeding (COB) are the popular tools to implement the participatory PPB approach. In PVS, available crop varieties are demonstrated in fields through farmers' leadership and in their management conditions. There are three steps in the PVS implementation: (1) variety collection, (2) experimentation in farmers' fields, and (3) scaling-out. The varieties to be considered in this process would be modern and farmers' varieties; thus, PVS allows farmers to choose appropriate varieties in their management. In COB, farmers get opportunities to select lines of their interest in the segregating materials (i.e., genetically non-fixed material) in their management condition. This strategy offers farmers wide selection opportunities with reference to their preferred traits, which is just opposite to that of conventional breeding where only a few selected lines are evaluated in farmers' fields. Moreover, some of the better lines selected from COB are further evaluated in larger plots in PVS. The adoption of PVS and COB in the national agricultural research stations takes time.

As these organizations are mainly guided in the top-down approach, some innovations need to be made by research stations for the adoption of the PPB approach. Some of the areas of innovation include appropriate models for implementing PPB in the target environments, variety registration, and stakeholders' participation. Another concern with research systems is the low volume of source seed available to seed producers. Agricultural research organizations could address this issue by increasing areas under source seed production by mobilizing their own networks, or by contracting out this facility to the private sector. However, more innovation is needed regarding cost–benefit analysis for source seed production through different channels. Understanding the cost–benefit situation would be helpful to develop appropriate source seed production schemes at local levels.

Government extension offices increase peoples' awareness toward improved seeds and associated production technologies through training, demonstration, interactions, etc. But the training content of these institutions is mainly focused on technical dimensions, neglecting entrepreneurship and institutional development. So, the entrepreneurship concept should be integrated in the training curricula of extension agencies, but these agencies should innovate how these concepts could be integrated considering the trainees' categories and the depth of knowledge required for them. Similarly, how to promote CSP on a wider scale is a common concern, and government extension agencies are in stress for not being reached in marginal areas because of their limited resources. In this context, developing mechanisms to strengthen CSP through NGO networks would solve this problem.

Seed regulatory agencies design policies for the seed quality control mechanism. The seed testing laboratories provide this facility, but they are mostly concentrated in urban areas, and owned by government agencies. In this context, the regulatory agencies should design appropriate policies to encourage people to establish seed laboratories at local levels. The concerning issue here is the development of incentive mechanisms to attract local people to set up seed laboratories at private levels. Similarly, it is important to promote a truthful labeling scheme in CSP because this approach substantially reduces seed certification costs and empower seed producers to assure their seed quality themselves. This approach does not mean that national seed laws should not monitor the seed quality issue in CSP. Several malpractices might happen in the process of implementing CSP, especially in the early phase of their development, because of poor implementation of seed law; for example, selling grain in the name of seed just by changing sacks. This problem could be addressed by formation of a local level (district) seed management committee involving seed producers, traders, and research and development agencies, and its mobilization to monitor seed quality in the market. Another strategy would be encouraging seed producers to diversify seed packaging size so that distributors do not have an opportunity to handle loose seeds.

Similarly, malpractice might happen while distributing seeds by development agencies in the name of humanitarian aids, just after conflict or disasters. These agencies normally choose seed distribution as a means of cooperation because seed is cheap, easy to handle, and applicable in the agrarian societies. Seed distribution, however, negatively affects CSP if it is not properly handled. For example, if relief agencies import the bulk of seeds from foreign countries without understanding their appropriateness in the targeted agroclimatic and socioeconomic conditions, CSP will be affected from two perspectives. First, seed producers could not sell the anticipated amount of seed in the market because consumers could receive seed free of cost from humanitarian agencies. Second, if the performance of distributed seed is not good it hampers the credibility of improved seed in the farm communities, and thus affects seed demand. To address these complexities, the seed regulatory body should design appropriate guidelines for seed-distributing agencies on the basis of seed security assessment in the targeted environment. A local government agency could coordinate among those associated with the seed subsector, which could promote public–private partnership to develop physical facilities in seed producer organizations.

11.3.2 Informal Stakeholders

The role of NGOs in empowering in technical and managerial aspects of seed production has been well established; however, NGOs should change their strategy of project implementation from input distribution to capacity building and networking (Table 11.2) as the failure of the input distribution approach has been already learned in establishing sustainable CSP. Similarly, NGOs could empower SPOs for improving their organizational governance, such as mechanisms of incentive systems, business plans, linkage, and participation. However, they need to innovate how they could reach marginal areas, and how the lessons learned by NGOs could be disseminated in the wider mass.

Table 11.2 Innovations with reference to problems across the informal stakeholders

Problems	Innovations
Nongovernment organizations	
Activities focused on materials supply	Activities focused in capacity building and networking
	Organize interactions in the leadership of seed production organizations (SPOs)/local clubs
	Development of modules to operationalize business plan in local language
Poor access in marginal areas	Mobilize trained/experienced seed producers to facilitation
Seed producer groups/cooperatives	
Low capital reserve and poor physical facilities	Appropriate modality for increasing business transactions such as increasing membership or contracting with smaller groups
	Develop good proposal for public–private partnership and contact service providers
Poor organizational governance	Appropriate incentive systems to poorer members
	Explore mechanism for realizing social incentives by SPO leaders

Farmer Groups and Cooperatives

There are two important concerns in seed producer groups and cooperatives: low capital reserve/poor physical facilities, and poor organizational governance. The first aspect could be addressed by increasing business transactions by motivating seed producers to sell the maximum proportion of seed in the market through the promotion of group saving schemes, training, and irrigation facilities. Another way to increase seed transaction volume at SPOs could be increasing their membership or contractual arrangement with farmers for seed production in nearby villages. Moreover, SPOs could also supply inputs that are complementary to seed production such as organic fertilizers, bio-pesticides, and agricultural tools. These activities could also help SPOs to maintain cash flow around the year and to minimize their management costs. The SPOs could also attract support from government agencies and NGOs to develop their physical facilities through a public–private partnership approach. For this, they need to develop convincing proposals demonstrating how they can contribute to the proposed project and the monitoring and evaluation system of the proposed projects.

Seed producer organizations are also poor in business plan development and incentive systems, and development of these systems is related to the education of their executives. However, most of the SPOs leaders have poor education. It is also difficult to bring educated members in as SPO executives as they could get better opportunities and therefore might not be willing to take on that challenging position. Even if they want to enter an SPO, other members of SPOs would feel it was risky to offer them leadership positions because of the possibility they would drop out in a short time period. In the third case, the existing leaders, even if their performance is poor, do not want to leave that position for lack of incentive, because these leaders take benefits from SPOs in indirect ways (e.g., improved social relationships). However, there would be three options to promote educated leaders in SPOs: (1) selecting higher-educated leaders from the existing members, (2) creating an opportunity for entry of educated leaders outside SPOs but from the same locality, and (3) providing a higher education opportunity for the existing leaders.

The implementation of the first two options is related to the development of incentive systems at SPOs. In this case, the incentive should be sought not only for potential educated leaders but also for existing leaders who have already been in a leadership position for a long time. The contribution of existing leaders needs to be respected by members because they handled the organization in difficult situations when other members were not interested in taking on the management responsibility. The SPOs might not be able to provide financial incentive to their leaders, especially those SPOs in the early developmental phase, considering their poor cash reserve. Rather, consideration of social benefits such as recognition as advisors or life members of SPOs would be useful to sort out this problem. The third option for developing educated leaders in SPOs is sending leaders for higher study. However, vocational training could be more appropriate for SPOs in the early phase of development because most of the leaders are middle aged.

11.4 Conclusion

Community seed production catalyzes innovations in both formal and informal systems to enhance food security in rural areas. This system is mainly owned by smallholder farmers in marginal areas and is promoted by development projects on a small scale. However, the lessons learned from the system should be institutionalized for its further expansion. These lessons are the adoption of participatory approach, dynamic leadership, entrepreneurship, and capacity building. However, the concerned stakeholders should be enabled to internalize the relationship of these lessons to the benefits they intend to get from seed production and marketing. This idea necessitates stakeholders' active engagement in the process of implementing CSP in farmer-led and other supported models. Once the players are engaged and interacting with each other, they internalize the key problems in the seed production and consumption chain and share lessons they have learned. Organizing joint monitoring visits at seed production demonstrations is a common practice for engaging participants in discussions, and these activities are mainly organized by development projects. However, organizing this event by SPOs would enhance their capacity and reduce the costs for organizing the event.

This book has measured the economic, environmental, and social performance of seed producers engaged in rice and wheat seed production and operating in a low institutional development and poor capacity. As already discussed, some lessons have been generated, and some approaches in measuring the performance of CSP have been validated. The most important approach is establishing a linkage between organizational governance and household-level economic benefits, opening avenues for extension agencies and policy makers to design their policies for strengthening CSP. However, lack of soil fertility data prevented relating governance indicators to environmental performance. So, future research should address this issue. Other important topics to be covered in future studies include the following:

- Modalities and cost–benefit analysis for source seed production schemes at local levels
- Roles of organizational structures in the performance of seed producer organizations
- Appropriate modalities for developing crop protection/insurance schemes in CSP

References

Bishaw Z, van Gastel AJG (2008) ICARDA's seed delivery approach in less favorable areas through village-based seed enterprises: conceptual and organizational issues. J New Seeds 9 (1):68–88

Eigu J, Waters-Bayer A (2005) Unlocking farmers' potential: institutionalizing farmers' participatory research and extension in southern Ethiopia. Project experience series 2. FARM-Africa, London

FAO (1998) Developing seed security strategies and programs for food security in developing countries. In: Proceedings of the international workshop on seed security for food security: contribution for the development of seed security strategies in disaster prone regions Florence, 30 Nov–1 Dec 1997
McGuire S, Sperling L (2011) The links between food security and seed security: facts and fictions that guide response. Dev Pract 21(4-5):493–508
McGuire S, Sperling L (2013) Making seed systems more resilient to stress. Glob Environ Change 23:644–653
Srinivas T, Bishaw Z, Rizvi J, Niane AA, Manan AR, Amegbeto K (2010) ICARDA's approach in seed delivery: technical performance and sustainability of village-based seed enterprises in Afghanistan. J New Seeds 11(2):138–163
Witcombe JR, Devkota KP, Joshi KD (2010) Linking community-based seed producers to markets for a sustainable seed supply system. J Exp Agric 46(4):425–437

Index

N.P. Khanal, K.L. Maharjan, *Community Seed Production Sustainability in Rice–Wheat Farming*, DOI 10.1007/978-4-431-55474-5

MIX
Papier aus verantwortungsvollen Quellen
Paper from responsible sources
FSC® C105338

FSC
www.fsc.org

If you have any concerns about our products,
you can contact us on
ProductSafety@springernature.com

In case Publisher is established outside the EU,
the EU authorized representative is:
Springer Nature Customer Service Center GmbH
Europaplatz 3, 69115 Heidelberg, Germany

Printed by Libri Plureos GmbH
in Hamburg, Germany